Made in
MAINE

Made in MAINE

From Home and Workshop, to Mill and Factory

Paul E. Rivard

Charleston London
History
PRESS

Published by The History Press
Charleston, SC 29403
www.historypress.net

Cover photography by Donald Bassett.

First published 2007

Manufactured in the United Kingdom

ISBN 978.1.59629.218.5

Library of Congress Cataloging-in-Publication Data

Rivard, Paul E.
Made in Maine : from home and worskshop to mill and factory / Paul E. Rivard.
p. cm.
Includes bibliographical references and index.
ISBN-13: 978-1-59629-218-5 (alk. paper)
1. Manufacturing industries--Maine. I. Title.
HD9727.M2R58 2007
338.09741--dc22
2006034635

Contents

Foreword

When I was growing up in Maine in the 1950s and '60s, textiles were still being made in Lewiston and Skowhegan, shoes in Auburn and Freeport and paper in Brewer and Old Town. The world of my youth was rooted in the last phase of an industrial revolution that had begun with the harnessing of waterpower and steam in mid-eighteenth-century England and spread across the Atlantic to the new United States in the 1780s and '90s.

Two centuries later the dominance of manufacturing has been replaced by a service-based economy driven by high-speed electronic communication. Yet we are all children of an industrial age in which men and women spent their lives working in factories and shops creating products on machines and with their own hands. In my case, my English immigrant grandfather, George W. Shettleworth, was an expert machinist who fashioned governors for steam engines, while my Danish immigrant grandfather Knud D. Knudsen worked as a pattern maker in the Portland Stove Foundry. These two men experienced ten-hour days, six-and-a-half-day work weeks and remained at their jobs into their seventies.

As we move into the Internet era, the Maine State Museum's Made in Maine exhibit and Paul Rivard's accompanying book of the same title become increasingly relevant in reminding both ourselves and our children of the not-so-distant past from which we came. Created by Rivard and his museum staff in the 1980s, this exhibit powerfully transports us all back to a variety of nineteenth- and early twentieth-century work and home settings. We can walk through a textile mill complete with its looms and gaze upon a parlor that could be waiting for our great-grandmother to return from her domestic chores.

Now, some two decades after the opening of Made in Maine, its chief architect, Paul Rivard, has given us an excellently researched and well-written book that provides both significant context and amplification for the exhibit. Following in the recent footsteps of his fine volume, *A New Order of Things*, this current publication is a compelling essay on the history of Maine's industrial development, which is enriched by the careful selection of many appropriate period illustrations. Rivard's book reminds us that the Industrial Revolution played a pivotal role in shaping the Maine we know today and that we are still fewer than six degrees of separation from that defining era.

Earle G. Shettleworth Jr.
State Historian
Augusta, Maine

Preface

The Made in Maine exhibit, occupying several levels and totaling more than seventeen thousand square feet, was opened in 1985 following nearly six years of planning and construction.

Since the exhibit medium is primarily visual, I argued many years ago that the museum should remember that if things did not work visually, they didn't work at all. It was understood that visitors would not come to the museum to read labels and text. As a consequence, it was understood that the raw materials for Made in Maine could never be shown by the exhibit alone. Handout papers and a book would be needed to augment the physical installation. Thus, this book was envisioned in the early 1980s as a supplement to the installation.

The research leading to this book was used in the building of the Made in Maine exhibit at the Maine State Museum. To a great degree, though, this exhibit emerged as a lifelong interest of mine and this book necessarily includes material developed personally over the past thirty years, including work at the Slater Mill, the American Textile History Museum and as a founder of the Society for Industrial Archeology. As an organizing concept for Made in Maine, the workplaces of Maine were divided into four: the home, workshop, mill/furnace and factory. This proved to be an overly simple strategy but it was clearly useful. Depending on the level of capitalization and marketing, all levels of manufacturing coexisted at the same time. Regardless of how they might have started out, most manufactures ended up as factories. For example, almost all Maine shoemaking ended up in factory environments even though many cordwainers had worked at home or produced custom work on a bespoke, itinerant basis. All kinds of manufacturing were undertaken in Maine at the same time, and this great variety makes analysis difficult.

On the evening before the exhibit opening, I went to the museum around 9:00 p.m. and turned on all the exhibit lights to have one last look. Years had been invested in this, and I spent more than one hour visiting the installation alone. When I was done, I thought to myself, "Paul, you can never do better than this." I have enjoyed many projects since that time but, to this day, surely I have never done better.

Acknowledgements

I left the Maine State Museum in 1991 and took the raw material for this book with me. I had written it on my electric typewriter and I thought the advance from the manual to the electric was a great improvement. For me, computers were still in the future.

I carted the text and notes with me as my life moved on to another stage and finally, thinking that it would never get produced, I made the terrible mistake of losing it between moves. It was gone, and it is here that I want to acknowledge the very critical role played by a friend and supporter of the Maine State Museum. I returned to Maine in the twenty-first century and was asked about the Made in Maine book. I confessed that it was gone and I was told, to my great surprise, that a copy had been saved. The old typewritten copy was sent to me and, for this, I remain forever grateful to Representative Neil Rolde of York. He sent me a copy in 2003 and with Neil's attention I was allowed to revisit this yet again. Now I put this on computer and begin to edit and shorten it.

Without Neil Rolde's action, nothing more could ever have happened.

Without the help of the museum staff between 1979 and 1985, it is unlikely that the exhibit or this book could have been done. To all of them should go my thanks and lifelong appreciation.

In 2006, I was told about The History Press by several individuals and I sent the revised manuscript to them for review. The editors there accepted the work and I am grateful to this press for their interest and support.

The photographs used for the exhibit and the text are most often in the collections of the Maine State Museum and the Maine Historic Preservation Commission. It is thanks to J.R. Phillips, director of the Maine State Museum, and his employee, Deanna Bonner-Ganter, that I have been allowed to use these images. My old friend Earle G. Shettleworth is director of the Maine Historic Preservation Commission and also state historian. With the help of his employee, Liz Trautman, I have been able to use the collection of the Historic Preservation to get all images scanned for the press.

This book uses my name and after many thousands of hours I am pleased, thankful and proud. But the list of people whose contributions have made this possible is very long indeed. My family—Suk Hi, my wife, and Sonya and Chris, our children—allowed me the time to research this work and now, years later, they have again helped me get the book done. Without their forbearance, little I have ever done would have been possible. Thanks to everyone who made this possible.

Introduction

The romance of Maine is celebrated today through greeting cards and calendars featuring the state's "over the river and through the woods" character. Rocky coastlines, babbling brooks, lighthouses and lobsters loom large in these images. Millions who cross the Kittery Bridge each year seek to take their vacations in places where grandma might have lived. Thus, the album of Maine history is filled with images of the north woods and the sea. Mariners and lumberjacks are colorful ancestors for us; the smell of pine tar and oakum provides the aroma for a romantic past.

The profits arising from Maine's timber resources, and the attraction of life building and sailing aboard ships, directed the energies of Maine people away from artisan pursuits. When compared to its two neighbors, it is found that in 1850 the state employed over 3,000 sawyers and over 2,200 ship carpenters, as compared to a total of only 178 in Vermont and 156 in New Hampshire. From 1634 when the *Pied Cow* brought sawmill parts to Maine, the state had been an emporium for timber. There were over 200 fishermen in the state, while only 187 in New Hampshire and none in Vermont. When the title "mariner" is added, it is clear that Maine provided 23,000 of her men for the forest and the sea—a statistic that made the state unique in northern New England and limited the workers devoted to artisan trades.

Pine forests, which supplied ship timber, tar, turpentine, pitch and other "naval stores," drew the first attention of Maine's earliest settlers. Agriculture and the abundant forests exerted a later influence over all of Maine's commercial and manufacturing development.

But, alas, historically Maine people did not live this way and families probably shared baked beans more often than lobster bisque. On their way to noted landmarks and natural environments, visitors to Maine probably ignore the shops, mills and factories they might pass. Perhaps the workers in these places realized how lucky they were, but it is more likely that the natural environment was frozen out of their thoughts by the demands of earning a living. Even when most Americans called themselves farmers, tilling the land of Maine was never easy. Whether sweeping away the dust of sawmills or textile factories, the forests and the sea were seldom places of leisure; more often they were places of work. To gain an appreciation of Maine we might dwell on the state's special properties, but to really see Maine people we must consider the work and the products that often defined them and occupied their lives.

Economic development has always been a principal concern of state government. Accordingly, it is not surprising that Maine retained the service of Boston geologist Charles T. Jackson to survey the state's mineral resources as early as 1836. Rocks and gemstones were assembled during this effort and then led to establishment of a "cabinet" at the State House in Augusta. This was the beginning of the Maine State Museum. By the 1860s, the state sought to identify its

This Bradley photograph shows two hand sawyers cutting boards using a traditional pit saw. The man working above lifted the saw after each stroke and guided it down a mark placed on the log earlier. The pitman below pulled the saw back down but suffered the misfortune of having the sawdust fall on him. *Courtesy Maine Historic Preservation Commission.*

waterpower rights, illustrating prior activity and success and luring business to make use of the many horsepower that stood unused. Three books detailing the "Water Power of Maine" were produced by the State between 1867 and 1869. These efforts did much to quantify the state's potential and to identify the activities illustrating most all Maine manufacturing studies.

This book concerns manufactures and, except for printing calendars and managing motels, it speaks of another Maine. Today, for example, a principal stop for many Maine visitors is the town of Camden on the mid-Maine coast. Tourists who climb nearby Mount Battie can see an overview of the whole town and region. Seasonal gift shops and restaurants abound in a renovated white-clapboard downtown street. Schooners move in and out of the bay, ferrying visitors aboard the "tall ships." This is an idyllic Maine view, and even in the 1860s, the scenery of Camden was described as "varied and beautiful." But this image does not fully reflect the reality of the nineteenth century, when the largest building in town, built in the 1860s, was the Knox Woolen Mill. With nearly five thousand inhabitants in the 1860s, the town was growing and attracting manufacturers. "Camden," it was reported, "has devised liberal things for those willing to assist in developing her manufacturing resources, by exempting from taxation the property invested, for a period of five or ten years." The woolen factory specialized in producing papermaking felts sent around the country even while efforts were focused at bringing papermaking to Camden itself. Other businesses located up and down Megunticook Stream including an anchor factory; two gristmills; a "cracker bakery"; a marble works producing tombstones; an oakum factory;

a machine shop producing capstans, pumps and other ship building components; "Farrar's Carriage Factory"; a powder mill; and various other factories. This was characteristic of all of Maine at this time: tourism would largely come later, but manufacturing was the present.

The vast majority of products made in Maine, and America, were generic at first. The names of specific makers may have mattered locally, but were otherwise unnecessary as the products joined many other commodities in a wider marketplace. Thus, even heavy machinery was plain and unnamed when it was intended primarily for use by the parent corporation. It was simply impractical to think of marking names on many products, including lumber, grain and many items of homemade clothing.

Identification of products by maker reminded too many of an older, custom production more typical of local consumption. When commodities were produced by many different hands and intended for a larger market, standardization was more important. Thus, a bushel had to be a bushel every time, regardless of product or of packing agent. A quart was necessarily always a quart. So, then, generic textiles were woven in various homes and distributed as the same thing: the weights and sizes of yarn were agreed-upon standards and all the cloth had to look and feel alike. There was little room for individual flights of fancy in this world. Home weavers and sawyers followed precisely the agreed-upon rules of commerce.

For the curators trying to identify products of any region or maker, attention to detail is essential. National and even international fashions influenced many products used locally and impacted all products intended for commercial sale. Fashions, and attention to changing styles (from furniture to clothing), help in dating most products. Regional preferences can also be seen and choices of materials and colors can sometimes be traced to individual makers, allowing for the attribution of products not otherwise marked.

Up through the early nineteenth century the branding of products was infrequent. Makers were rarely bold enough to be individualistic. Most manufacturers were conformists who did not buck the pressure of fashion or design. But there were exceptions arising from the exploitation of prominent names or countries of origin. Since European textiles were more highly prized than those made in America, it was not uncommon for deceptive labels to appear in marketing these items, and many local products were equipped with tags showing they were made in England or France. Some product lines and manufacturing districts became identified with high quality and producers of these items were anxious to protect themselves as makers of "name brands," even when the production was widespread and similar. Thus the Scotch-Irish producers of quality linens in Londonderry, New Hampshire, sought to safeguard their market niche by controlling use of the place name as early as the 1720s.

In many cases services were needed near the workplace. This group included carding mills, clothiers' shops, gristmills and blacksmith shops. But most local goods found their way into a larger marketplace. Broomcorn came from Illinois, and clay from New Jersey. At the same time, Maine shoes were sold throughout the Southern United States, while spinning wheels went to the Great Plains. We enjoy the concept of self-sufficiency in rural America. In Maine, this was often a choice, not an absolute need.

A change in marketing strategy in the nineteenth century led to the growing use of name brands and an effort to protect the "genuine article." The evolution of this strategy was unequal, but increasingly manufacturers were anxious to identify their work. While many generic products remained, by mid-century products ranging from carding machines to sewing machines were marked with a maker's name and name plates appeared on a myriad of plows, sleighs, looms and even spinning wheels. Here, the task of the curator was made easier in a world of increasingly limited guesswork.

Whether they were generic or branded, Maine products were influenced by, and participated in, commerce that was regional, national or even international. Maine towns were very often rural but rarely isolated. Seaborne and river traffic brought almost any goods to the immediate areas of settlement. Fashion developments were usually well known and the production of everything from sofas to dresses was influenced accordingly. The role of the local shopkeeper was often critical and it was in this emporium that goods were distributed, products bartered and raw materials provided to shops and households seeking to pay bills and purchase other goods "from away."

Some services were meant principally to serve households or a strictly local community. In the modern world the home is commonly viewed as a "castle," or a place of refuge where families can escape from the rigors of the outside, work-a-day world. Through the nineteenth century, however, the "home" was often at the center of work life as well as a living space for extended families and hired hands. From here, family members worked or left for work. It is important to recognize that living spaces were also working spaces through the mid-nineteenth century. The specialization of products that characterized commercial and industrial development was very slow to replace the tradition of homemade products for domestic and local consumption. Furthermore, it was not until the twentieth century that minimum wage and child labor laws ended the exploitation of in-house labor for broader commercial purposes. Maine's houses, in 1820, might be seen as many islands of manufacturing activity, places of work that were workshops for both household goods and income production. Aside from sawmills, houses were actually the first places of manufacturing in the state. The most important domestic products were textiles produced in excess of the need and, in the second half of the nineteenth century, spinning and weaving were followed by hand- and then machine-sewing of garments.

Spinning and weaving actually survived in Maine a great deal longer than might be expected, given the rise of both woolen and cotton factories. Wool spinning machines were produced, to be sure, but the diary of Persis Sibley Andrews of Dixfield, in August of 1845, is particularly revealing:

> *We have a loom set up in our back kitchen and I am weaving my web of wool flannel myself. Father gave me a bag of wool. Costella has spun the filling—I hired a woman to spin the warp, & I got 18 yds, of nice cloth pretty cheap. Mrs. W. has a carpet to weave. It is not common now to make cloth in families, but we keep each of us a hired girl, & both Mrs. W. & myself enjoying better health than usual—we have tho't to do a little something to convince the neighbors that we sho'd be smart women if we were not nearly always sick.*

Household crafts were sustained by an abundance of raw materials and a frequent lack of specialized skills or tools. Enterprising merchants devised numerous ways to introduce labor-intensive piecework into the less structured home environment. One shopkeeper in Westbrook,

Only a relatively small number of mills in Maine were operated by tide power. This tide mill in South Harpswell was one example. *Courtesy Maine Historic Preservation Commission.*

for example, cut paper patterns and sent them out to be pasted into bags in households. This was a beneficial commercial pattern that extended directly from the home kitchen to the marketplace, with the shopkeeper as middleman.

When most men preferred to call themselves "farmers," the house was on land that was growing crops and the nearby barn was home to the animals, the carriages and the tools. Additional work in shops, mills and even factories supplemented the farmer's life. Women sustained the households and managed the drudgery of many chores ranging from rearing children to preparing meals. They sustained households that included the family and hired men and women and, permitting no idle time, they worked to produce goods for the outside world. Prior to textile factories starting in the 1840s, the most common outside job facing young women was that of "domestic," serving the needs of other large families and, again, staying and working at home. This was a job choice they avoided when possible. Girls often preferred the cash and independence first offered by the factory and their in-house employers were not pleased with this competition.

The commercial exploitation of labor in the home tended to make the activities of domestic manufacture more specialized than they had been when the primary concerns focused on family and neighborhood needs. In this regard the home environment sometimes came to resemble the workshop. Martha Ballard's "loom roome" and Elizabeth Smith's "spinning and weaving room" were very similar to the general repair workshops attendant to most farms and, when there was a consistent use of space for commercial purposes, these were, in fact, true workshops. It was probably the tradition of textile manufacture in the home and the direct influence of women that fostered these activities in the home. Meanwhile, most other craft

This sawmill in Caratunk sports a simple dam of logs and gravel. *Courtesy Maine Historic Preservation Commission.*

manufactures, whether by convention or simply because they created noise and mess, saw the tools of their trades assembled elsewhere.

Despite its utility to manufactures, the home remained a living area. Tools, whether spinning wheels or sewing machines, were tolerated intrusions. While looms might be set up temporarily in the kitchen, the artisan workshop was a place intended as a home for the tools.

Craft specialization defined the workshop and distinguished its products from those of the home. At first, skill was relatively more important than machinery and so, in the eighteenth and early nineteenth centuries, the existence of craft workshops presupposed the involvement of skilled craftsmen. But these crafts were often in short supply in colonial America and this shortage, together with the absence of craft guilds and tradesmen's organizations, provided a foundation in Maine for the grass-roots emergence of workshops in the agricultural population. In eighteenth-century Maine, for example, the density of population was seldom sufficient to encourage the existence of craft workshops with full-time workers. Commerce, not local manufacturing, supplied most wants for things like pewter, pottery, silver, tin products and furniture.

Craft manufactures in Maine between 1760 and 1800 must be characterized as rising mostly from workmen, sometimes known as farmer-artisans. While the Maine market could not itself foster extensive craft manufacturing, the products of the soil were seldom adequate to supply the family's entire income. Conversely, skilled craftsmen often lacked sufficient market demand to follow their trades on a full-time basis. This was the case of Moses Davis of Edgecomb, who was both a farmer and cabinetmaker. In addition to the work of planting, plowing, harrowing and harvesting, Davis reported on September 11, 1783, that he "workt in Shop on Capt. Dutton's Chest." He also was engaged "making a Great Plow," "finishing a chest for John Place" and making "a standing stool for Mrs. Decket." Davis helped in barn raisings, "workt on joiner work for vessel," framed houses and built the custom cabinetry for a kitchen in Wiscasset.

The work of Moses Davis presents an appealing picture of a Yankee grass-roots craftsmanship but, alas, this had little to do with real life in Maine. The work of farmers was more commonly found in the production of shingles, clapboards and barrel staves. Meanwhile, many artisan workshops were a result of regional, even international, commerce, and the influence of the forest and the sea tended to divert energy toward these activities.

In terms of business development, the lowest stage or organization was that of custom, or bespoke, manufacture. At this level the craftsman was engaged in producing commodities on order. Examples of this include a blacksmith shoeing a horse, a shoemaker fitting a pair of boots, a saddler making a saddle for the family horse and Moses Davis making "a chest for Mr. Silaway." This tradition of craftwork has continued to the present day and these manufacturers are often known today as jobbers. These heirs to the blacksmith shop are today the managers of service stations. This work required relatively little capital, little risk-taking and no merchandising. Conversely, there was little job security.

Many workshop crafts evolved rather quickly into larger-scale production of "sale work." At this level of business organization, commodities were produced at the pace of men and tools, and the products were marketed through wider distribution and marketing. "Name brands" emerged here. In 1820 a great number of Maine's 1,768 "mechanic workshops" were engaged in bespoke manufactures "immediately connected with and indispensable to, the ordinary pursuits of agriculture." A rapid growth in these manufacturing activities occurred in the early nineteenth century and these were encouraged by a preexisting network of rivers and harbors that had at first retarded local production in the face of commerce. In the nineteenth century this avenue to the work, long exploited by sawmills, became the great impetus to the employment of artisans and craftsmen, and only a few bespoke manufactures endured on a local level.

The waterpower of Maine provided power for many mills and, later, factories. Here men and women engaged in agriculture could earn extra money and young men with little or no household property could find employment. Sawmills, carding mills and gristmills provided only seasonal employment anyway and this well suited the life of rural America not yet accustomed to wage-earning schedules. Meanwhile, a range of products emerged from the home and the workshops, the work of so-called farmer-craftsmen.

Some 2,200 square miles of Maine is composed of lakes; over 1,500 bodies of water have been named. The vast majority of these lakes drain into and form the headwaters for the state's major river systems: over 300 lakes and ponds fall within the Kennebec watershed, while a whopping 467 bodies of water drain into the Penobscot. Today these inland waterways contribute to one of the state's largest industries—recreation. Through the end of the nineteenth century the term "vacationland" was not common and this drainage of water was instead viewed as a very valuable resource that was exploited by the construction of many thousands of mills and factories. Contributing to the value of this resource was not just the quantity of water but also its height above sea level. Moosehead Lake is some 1,000 feet above sea level, Richardson Lake is 1,400 and Rangley Lake over 5,000. This provides for a considerable descent and a large opportunity for the use of power.

Mill, furnace and tanning industries were consolidated in factory organizations during the nineteenth century and their functions were embraced within the broader commercial orientation of consumer products. Therefore, sawmills tended toward the manufacture of doors, sash and blinds while gristmills were associated with bake houses. Casting furnaces were the makers of stoves, architectural trim and fencing while tanneries retained their long association with the manufacture of boots and shoes.

A survey of Maine's waterpower potential that was prepared by the State in 1869 counted over three thousand "mill privileges," about half of which were not used. The report, advocating the use of this power, noted that:

> *The 1,229,200,000,000 cubic feet, more or less, of water annually delivered by our rivers, fall therefore on their age to the sea through the mean distance of 600 feet, and in their descent yield a gross power of 4,429 horse for each foot of fall. This being multiplied by the total average fall in feet, gives 2,656,200 horse-powers gross, which are equivalent to the working energy of over 34,000,000 men, laboring without intermission from year's end to year's end.*

In layman's terms, "vacationland" offered a lot of power. But extracting this power required dams, sluiceways, water wheels and turbines. From the 1630s to the 1830s, the technology of these wheels changed very little. Today, weekend explorers paddle through Maine waters in rented canoes while fly fishermen wade along streams in a quest to outsmart animals with a brain size less than a pea. Up to the mid-nineteenth century, however, many of these explorers were estimating waterpower for factories or millwrights, tapping the power of rivers and streams to drive mills. Water wheels were the prime movers and wood was the principal material until at least the late 1830s. For more than two centuries, trees were fashioned into water wheels, which captured waterpower and relayed it to gears and shafting to run the machinery of mills and factories.

While the iron and steel of foundries was destined to replace wood, there were essentially three types of wooden water wheels used up to nearly 1840: undershot, overshot and breast. Of these the easiest to build and manage and the most common was the undershot type, which operated through the force of running water striking the floats on the lower side of the wheel. Some were certainly used to drive tide mills, which operated by capturing the incoming tide and releasing the water onto the floats as the tide grew lower. There were at least forty of these as late as the 1860s. But most were, in fact, used as normal wheels along most of Maine's rivers and streams and known as flutter wheels. If water was already flowing down toward the sea, construction needs and cost were relatively small. Most sawmills, gristmills and later carding mills used these simple, if inefficient, wheels. Many sites were located on very marginal outlets and in places where the waterpower was pathetically small. Early photographs of mill sites often show simple buildings perched precariously on natural outcroppings along small rivers and streams. These mills appear to have invested very little on the construction of sluiceways and control gates and they were regularly swept away by freshets. As early as 1749 this fact in New England was noted in Douglass's *Summary*: "New England abounds in saw mills of cheap and slight work...These mills mostly stand upon small streams, because cheap fitted...[but] they do not afford water sufficient to drive the wheel in Summer."

Over a century later one mill in Wales, Maine, was still described as "on a brook, where some little sawing is done, during the spring and snow melting, and during the season of freshets."

It would be a mistake, though, to conclude that many Maine mill sites were this wasteful and there is evidence of the opposite. The mills built from the 1830s to 1860s were particularly conscious of the need to conserve waterpower through careful engineering.

Overshot wheels were rarely used in Maine. These wheels used water falling on the upper floats, and they were appropriate where there was relatively little water that descended a great distance. In 1860 there were still sixteen of these enumerated in Maine, the largest some thirty feet in diameter that was used to drive a diverse group of machinery at Shaker Mill in Poland as part of the Sabbathday Lake community. A sixteen-foot overshot wheel was used in Verona to run an up-down saw that, it was claimed, cut some two thousand board feet of lumber in twelve hours. These wheels were not easily replaced by iron turbines and, as a consequence, they often lasted well into the age of photography and are seen as the motive power of "quaint" mills, including one example in Moscow, Maine.

Both flutter (undershot) and overshot wheels were relatively simple, with an average running speed of only ten revolutions per minute for most undershot wheels. The needs of textile factories, however, were much greater and the engineering of water systems for textiles was much more involved and expensive. When textile mill villages and factory towns were first laid out, the investors planned dams and sluiceways to deliver water to each site and water wheels to capture this power and send it upstairs to the many needed machines using a system of gears, pulleys and shafting. Often, the wooden wheels needed for this task were breast, which operated with the weight of the water instead of its velocity or movement. This required that the wheel be fitted

A few residents of the Pepperell Manufacturing Company's tenements pose outside of their boardinghouse in Biddeford around 1875. This building was four stories high, 420 feet long and contained 256 rooms. *Courtesy Maine Historic Preservation Commission.*

carefully and tightly within breastwork under the wheel. Gates controlled the flow of water from above the dam and brought water directly to the wheel, where the weight of water on a fraction of the wheel's circumference caused it to turn. Water was controlled and little was wasted as factories turned with gearing and shafting leading upward from the breast wheel. A great deal of power was captured, but a great deal of money was invested, often by out-of-state investors. Virtually all textile mills in Maine prior to the late 1840s operated this way.

Over a century later the use of very limited waterpower was still common in Maine. In the nineteenth century, however, the use of waterpower reached enormous proportions in the operation of textile factories. Under the influence of major merchant capitalization, the management of waterpower emerged as a most important aspect of factory building that determined both geographical location and the pattern of growth. Large New England manufacturing cities such as Lowell and Lawrence, Massachusetts; Manchester, New Hampshire; and Biddeford and Lewiston, Maine, were dependent on securing and controlling adequate waterpower. As part of a general effort to improve water wheel efficiency came the iron water wheel of more scientific construction: turbines.

The first turbines in New England began to replace breast wheels in the early 1840s and within a very short time the U.S. Patent Office issued more than three hundred patents for various turbine designs and improvements. At first, the principal users of these new iron wheels were textile factories. But the advance of foundry technology worked to make these new wheels products of the foundry, not the millwright. Turbine wheels grew cheaper to purchase and wooden wheels were very soon displaced, even at much smaller mill sites. Turbines were easy to install, efficient and fast moving. Wooden wheels were considered obsolete and old-fashioned. By mid-century these turbines were the prime movers of choice for most Maine factories.

As a cost savings, investors often avoided the major waterpower sites of the bigger rivers. But the needs of factories brought with them the need for investors, often lured from out of state, and the exploitation of powers requiring dams and sluiceways to run a whole city built of factories. Up through the early nineteenth century, it had been true that urban areas had grown up at the seats and crossroads of commerce. Sometimes, as in the case of lumber cutting, this commerce arose from local raw materials sent away in return for other commodities or money. Trade created these cities and made many merchants wealthy along the way. Until the birth of textile manufacturing, few merchants saw clearly that manufactures could be a competing source of investment and riches.

As elsewhere, Maine's factory towns included power and a way to tap it, factories and housing for factory operatives and their families. This was the story found in Biddeford/Saco, Lewiston and countless other factory cities—urban areas with manufacturing jobs. Factories may have created and offered work, but rarely were they also places of refuge from work. Lewiston was a place you took vacations *from*, not vacations *to*. There were few factories among the attractions of vacationland, and visitors to Camden have usually visited the harbor, not the Woolen Mill. To many visitors, lighthouses are the "real" Maine and factories are not. Thus, lighthouse keepers get good press and factory workers do not. As a direct consequence, Maine's working people are not seen by tourists crossing the Kittery Bridge, and it is this body of workers that is central to this book.

I.

The Family of Workers

Most of the manufactures of the State are conducted in small establishments, and many of them in private families, a part of whose time is employed in agriculture.

Moses Greenleaf, 1829

When Maine became a state in 1820, its population of over 300,000 persons occupied some 32,127 dwelling houses. Beyond these statistics, though, it has always been extremely difficult to look into these rooms and to see the labor of the families at work there. We draw clear distinctions now between the home and the workplace, but in 1800 those distinctions were less obvious. Household prosperity, and sometimes even household survival, depended on domestic trade, which extended from the household to the store, to the neighborhood and even the region. Homemade goods were found everywhere and family enterprises to make them were well known and ubiquitous. With the possible exception of sailing and sawmilling, these were the first workplaces in Maine and they existed throughout the state.

The work of whole families is often hidden by the prevailing practice of listing all accounts and debts under the name of the "head of household." Little cash ever changed hands and, as a consequence, it is difficult to quantify the working life of the family members and easy to understand why many saw household labor as unrewarded drudgery. Work in homes was rarely acknowledged, even when these tasks left the house. Commonly, the work a family member performed was not recognized, especially when income was used to offset debts or settle accounts. But clearly in the eighteenth and early nineteenth centuries, Maine homes were places of both work and living. Product specialization that supported domestic and local trade was notable here too. With no minimum wage or child labor laws, Maine's 32,127 houses might be seen in 1820 as many islands of manufacturing activity. For generations, it was here that a very large quantity of manufactured goods was made in families.

In reporting our history, we favor romance over reality. Thus, lighthouses and tall ships are forever more popular subjects than overcrowded household and rural poverty—characteristics we now associate with child labor and factory living conditions. But to understand the house as a place of work we must peel back this romance to reveal a lack of privacy we seldom suspect. If we use Gardiner as an example of home life at mid-century, we notice that there were some two hundred families in town in excess of the dwellings. A total of thirteen hundred individuals (20 percent of the population) lived with other families. Few households could, in fact, be called single-family units. Living spaces and the work places were not distinct. Hired hands—craftsmen

Hand spinning with a typical wool wheel was depicted in this 1902 image for Center Lovell. *Courtesy Maine State Museum.*

and farmers—often lived with the owner's family and tramped from the fields, and through the house, to their small rooms. In 1850 William Day, a Gardiner blacksmith, lived with his wife, a daughter and granddaughter, as well as Sarah Maxey, the wife of a local mariner. Nearby, a shoemaker, James Robinson, provided room for three boarders, along with his wife and three children. James and Joan Reynolds shared space with William Rafter, Daniel Hern and Amasa Davis. It may seem unusual today, but these arrangements were common in 1850.

The Maine home was seldom a sanctuary in the mid-nineteenth century. At the same time, Maine communities were not isolated enough to warrant a description of "frontier subsistence." Commerce was general and the concept of self-sufficiency was largely a romantic illusion. Homemade crafts were sustained by an abundance of raw materials, a lack of needed skills or tools and local merchants who enjoyed the benefits of an in-home labor pool. Commercial networks arose not from "Yankee ingenuity," but from the family's desire to increase its income. For many farm families, manufactures resulted from part-time activities that included trimming staves, heading nails and making brooms, ox yokes, shoe pegs, caps, bedding and numerous other commodities. Homemade products like these could be used as a substitute for money in a cash-poor local economy. In Skowhegan, for instance, Eli Weston made plows, Albion Nay made ox yokes, Ebenezer Wentworth spinning wheels and sap buckets, while John White produced brooms. All these men were farmers, but they were also small-scale manufacturers and, to some extent, merchants.

In Westbrook, one shopkeeper cut out paper patterns, which were then sent out to neighborhood houses to be pasted into bags. This was a beneficial strategy that extended directly from the home kitchen to the marketplace and the shopkeeper. The incentive for this production came from the desire to increase family income, the shopkeeper's need to provide customers a means to exploit their labor as a medium of exchange and purchasing power and the use of relatively abundant raw materials and home labor. For these reasons, domestic-commercial manufactures became a general pattern of local trade.

Numerous commodities were produced in the Maine home (or as a byproduct of home life) but all were strictly secondary to the manufacture of textiles. This aspect of domestic manufacture was so omnipresent that the term "domestic manufacture" was commonly held in the eighteenth and much of the nineteenth centuries to mean the spinning of yarn, weaving of cloth and sewing of clothing. All other domestic products were considered unimportant. The home was the principal workshop for making textiles prior to textile factories or ready-to-wear clothing. While many other aspects of craft manufacture led to workshop industries, textile production went from the home to the factory and clothing production in the home became ready-to-wear manufactures.

Homes were often supported by local mills for carding and fulling and by professional cloth finishers and tailors. But there is a common misconception suggesting that cloth production moved from the home to the factory as machine processes decisively reorganized these manufactures. It is thought that in such a move, homes simply could not compete with machines. But rather, the evolution of these practices required almost a century to unfold. The world of machinery was redefined continuously to form an ever-changing jigsaw of supporting parts. While machine innovations were introduced as early as the mid-eighteenth century, for example, spinning wheel improvements were patented for another century. This slow transition sustained a significant role for many Maine homes.

Textile production depended on its need for labor more than the need for raw materials. The medieval catalogue of fibers—wool, flax, cotton and silk—was not found to occur naturally in the New World. Rather, these were brought to North America from Europe along with the first settlers. The Maine wilderness and sea added nothing. Homespun was initially produced because it served local needs and used the labor available in households. At first, this was yet another avenue for domestic income, but eventually this was a path leading to Maine's largest factories. Although cotton and silk were well known as imports, the fibers most commonly used in Maine were flax and wool. Products of these fibers were very often produced on many Maine farms.

Growing and preparing flax was labor-intensive and required skill and patience. On the other hand, linen cloth was durable, and the raw material could be raised and prepared by rural Maine households and farms. Predictably, linen remained the most solidly established and widely distributed of New England domestic manufacture until overwhelmed by cotton fibers and yarn after 1810.

Like flax, wool was not native to Maine or New England. Ships carried not only flaxseed but also the stock needed to launch herds. The early catalogue of necessary animals included sheep. New England was inexorably becoming a clone of old England as raw materials and skills perfected in the mother country were relocated one by one to the colonies. Quality wool yarn relied on proper shearing of the fleece and sorting fibers by length and coarseness, and these were often tasks reserved for trained specialists.

Flax growing and sheep husbandry depended on the skill and work of farmers. But, once produced, the first task of the home was spinning these fibers into yarn. This work might seem romantic today, but spinning was a specialized task and it was hard work demanding experience and skill. Although men had been involved in agriculture and husbandry, the production of yarn and cloth were widely seen as the work of women and this may help account for the long survival of hand spinning.

Spinning was essentially a process of twisting to produce friction between individual fibers and holding them together to form a continuous yarn. Despite this, the spinning of linen and woolen yarns was different and each depended on the use of different tools. The wool wheel, sometimes known as the "walking wheel" or "great wheel," was the first to be introduced. It was the simplest, and it continued as the longest-lasting wheel in home use. Wool is relatively short, curly and barbed, so in preparing it for spinning it was commonly teased into a long strand of common size and consistency. The process of hand carding involved the use of two flat cards composed of steel wires set in a leather base. By working these cards against one another, a rolag of teased wool could be formed and this was the raw material for the spinning wheel. These rolags were the raw material for spinning and producing these by hand probably took longer than the actual spinning itself. Around 1800, mills and machines helped home spinners to produce these strands. Thus, the machine age fostered work in family homes long before it fostered factories. Carding mills gave home spinners a new lease on life, even while they provided the first machine installment toward the woolen mill. Here, the leather and wire card clothing was attached to circular drums and the process of teasing the fibers became relatively continuous. The carding machine delivered rolags in lengths of two feet (the width of the machines), and this was about right for hand spinning at the walking wheel.

Maine can boast one of the first operational wool carding machines used in America, established by Samuel Mayall in Gray around 1791. Carding technology in the province began here and by 1810 there were 75 carding machines producing rolags for spinning of over 450,000 pounds of wool each year. By 1820 there were 210 of these machines and the number was still growing while homes were assisted and encouraged to spin woolen yarn. Not surprisingly, in 1807 Hallowell's Martha Ballard sent her son, Cyrus, to the "masheen" with four pounds of wool.

The wool wheel used a large diameter driving wheel to give speed to a straight spindle. This speed was controlled while feeding in the carded rolag stock using just one hand. As the fibers fell from the end of the spindle, the fibers were twisted into a yarn. Rolags supplied from the carding machine were far more uniform than those produced using hand cards, and this permitted hand spinners to produce more yarn, and more uniform yarn, than ever before. In addition, hand spinners gained the use of another patent in 1808—the Amos Miner Patent Accelerating Head. This simple device offered a miniature pulley system that greatly increased spindle speed. Armed with machine-carded rolags and the Miner head, Maine hand spinners could now produce as many as three to four thousand yards of woolen yarn per day!

The spinning of flax involved a very different process. Here, the Saxony wheel, and many other similar devices, used a bobbin and flyer to both twist fibers into yarn and wind these on a bobbin. This was impractical for wool, but it was the standard tool for spinning yarns from flax and the cornerstone component of spinning machines designed for cotton beginning in the

1760s. Long flax fibers were fed to the bobbin and flyer by a seated spinner who drew fibers from the distaff above the wheel. There was no machinery developed for flax, no adapted version of the carding machine and no machinery at all to move this process from the home to the mill or factory. The use of linen yarn was thus doomed to near extinction in an age dominated by cotton. Because linen was strong and was a homegrown and spun product, its production ended slowly in Maine. But there was no factory stage in linen production, and by 1840 flax processes were essentially over.

Spinners produced only yarns, component raw materials of cloth itself. But spinning took more time than most weaving and more spinners were needed in Maine homes than weavers. Since spinning was the bottleneck in cloth production, and the tools were smaller and less expensive, nearly all Maine homes included spinning wheels by 1810 while only a fraction had looms. But ahead was the weaving of cloth itself. This was destined to be a factory-made product, but power-weaving equipment followed machine spinning in America by a generation. The gradual perfection of machine technology fostered, rather than threatened, home weaving. Again, family enterprise was sustained and rewarded. The 1790s to the 1820s proved to be the golden age of home weaving.

While spinning wheels were relatively portable, looms to weave cloth demanded significant house space and, often, a room of their own. Accordingly looms were usually the property of those with houses offering space to store and use them. The owners were not necessarily the actual users of these tools. In the same way that plows were owned by those with barns, and bedsteads owned by those with houses, weaving equipment, and often smaller tools, were stored, used and owned in households, even though farmers might give the work to temporary hired hands of boarders. Thus, Moses Davis of Edgecomb was annoyed in 1782 because his mowing "was hindered fixing a Loom & warping bars for Ellis Moore to weave," and Martha Ballard of Hallowell noted in her diary for May 19, 1787, that her son Cyrus "brought the bars, and other utensils for weaving, home" and in 1791 that "Mrs. Livermore put a web into my loom."

Although many weavers were men, weaving, like spinning, was largely considered women's work. A result was a local and regional commerce in yarns and woven goods. Domestic hand weaving illustrated well the commercial nature of the home as a workplace. Contemporary weaving demonstrations, common to historical museums, tend to reinforce a rustic view of textiles and it is easy to imagine that clothing was only slightly finer than a medium-weight rug. These homespun textiles accentuate an antiquarian theme but they do not reflect the quality standards of eighteenth- or even early nineteenth-century clothing. Wardrobes were certainly small when textiles were scarce and expensive. But our ancestors were not dressed in ragbags. Highly tailored and stylish garments were prized and accorded a top place in many household inventories well into the nineteenth century.

The 1810 Census of the Province of Maine provides a statistical glimpse of domestic manufacture. Over 16,000 looms were counted that year and the towns of Topsham, Lisbon, Lewiston and Bowdoin even enumerated the distribution of these looms among the households of the community and counted the yards of cloth produced. Of the 211 households in Topsham, for instance, there were 116 looms, while over 150 households reported the manufacture of cloth. This total was possible only with the use of hired help and borrowed tools. It was estimated that something over two million yards of cloth were produced in Maine homes. This weaving was the

This photograph, taken around 1875, shows a Maine farmhouse in the Androscoggin County area. Houses like this were often home to many work activities, while the attached buildings and American-style barn housed the animals, their feed, numerous other "dirty" jobs and agricultural tools. *Courtesy Maine Historic Preservation Commission.*

most widespread, and valuable, economic activity of the province, and almost one million yards of cloth in excess of the actual need was produced.

Linen, for all its strength and durability, saw only limited use in making clothing. It was wool, not flax, that kept people warm in winter and produced the most useful clothing, even in summer. Revolutionary War soldiers on both sides wore uniforms made of wool and, because this fiber was favored in making the best clothing, woolen garments endured well into an age of machines and factories.

The making of quality woolen goods was more complicated than might be suspected. While there was essentially only one kind of flax plant, there were many breeds of sheep. Accordingly, European countries long before had developed sophisticated methods of woolen manufacture and sustained a merchant network providing a steady supply of wool fibers of all grades. Homespun yarn and cloth confronted some very high expectations and, indeed, most products made in the colonies or in family homes were seen as second-rate. It was virtually impossible to compete with imported European products unless their true manufacture was disguised, or without the work of a professional clothier and the use of waterpower to drive the fulling hammers (stocks). Despite this handicap, it was estimated that in 1810 Maine households produced some 450,000 yards of woolen cloth. In 1820 a full two-thirds of fabrics made in America were "made in families."

Shearing sheep and sorting fibers were tasks left to experienced workers, seldom to local farmers. The principal artisan tasks and mill processes followed spinning and weaving. Through the "fulling" and finishing of woven cloth, these processes were essential to the production of quality woolen goods. Coming directly from the loom, woolen cloth was coarse, and at this stage it was in fact a flannel that sometimes resembled burlap. The best eighteenth- and early nineteenth-century clothing was made of cloth that had been felted and sheared so as to thicken the cloth and present a uniform nap of fibers. These actions rendered the warp and weft yarns indistinct and improved both appearance and insulating properties. These processes were known as "fulling" (a process of felting) and "finishing" (a series of processes that includes napping, shearing and pressing). All of these activities, in addition to dyeing, were found in the fulling mill/clothier's shop. The appearance and distribution of fulling mills documents the rise of woolen manufacture in the state through the late eighteenth century up to the 1830s. Thus, these mills can be traced to the seventeenth century even though it was not until after 1760 that the services of these mills and clothiers were more than marginally available to Maine weavers.

An 1815 American publication, *The Dyer's Companion*, provides some useful advice to home weavers and the cloth:

> *To make fine Broadcloth, take your wool and sort it carefully; take the shortest and finest of the wool, leaving no coarse locks with it; then break the wool all together, and card it into rolls by one person or machine, then spin well the filling cross-handed; give it a good twist, but not so as to be wirey; let it all be twisted alike, and spun by one person, then let it be well wove, with the threads closed together, but not too hard. Then take the long wool, and have it combed into worsted; have it well spun, twisted well, and wove firm. Thin cloth depend on the twisting and firm weaving but the other, after it is well manufactured, depends on the fulling to close and make it firm, and on the dressing for beauty. The cloth if well manufactured, well fulled and well dyed and dressed, will appear equal to any imported cloths; but if not well manufactured, it will not be handsome.*

The Maine home was never self-sufficient in producing woolen cloth. It was not until after 1760 that the services of fulling and finishing were more than marginally available to weavers in Maine, and throughout the whole period, including most of the eighteenth century, the capacity of Maine homes to produce quality woolens was limited. The appearance of fulling and finishing mills does help to document the otherwise almost invisible rise in woolen production. Judging from this evidence, the period of substantial growth in domestic manufactures was the early decades of the nineteenth century. Although one of Maine's first documented fulling mills was erected by Jonathan True of North Yarmouth in 1760, as late as 1810 there were only 59 fulling mills in the province, and this number grew to only 149 by the time of statehood. The skill level may not have been uniformly high, and this may explain why Jonathan True was only a part-time clothier.

The basic task of the fulling mill was felting fabric. The use of special solutions and the mingling of fibers with rollers or hammers (falling stocks) produced a tangled, compact mass that served to shrink and thicken the cloth. The clothier then raised the nap on this cloth with teasels and brushes and sheared the surface to produce a smooth and elegant finish. The finished cloth was expensive and almost never would have been produced strictly at home.

This image from Center Lovell shows the winding of yarn from a skein winder to a bobbin in preparation for making a loom warp. *Courtesy Maine State Museum*.

The Carleton fulling mill of Alna provides a glimpse of family woolen production and fulling. Some 3,600 yards of woolen cloth were treated here in 1809. This sounds like a lot of work, but the cloth belonged to 218 different customers, suggesting that the home production in Maine was widespread but not intense, with household quantities that rarely exceeded twenty yards, and often much less. This fulling and finishing was not intended for regional commerce, but rather for home and neighborhood use. The cost of actual weaving in 1809 was around $0.08 per yard but the cost of fulling and dressing amounted to nearly $0.23 per yard with a value of finished cloth around $1.25 per yard—indeed the most expensive cloth produced in the Maine home.

Wool and flax were both products of the New England farm and it is entirely logical that efforts would be made to turn these fibers into cloth and, ultimately, clothing. The missing fiber, cotton, was also important to Maine, particularly in the nineteenth century. This fiber was not raised in New England and, consequently, there were no cotton spinning wheels in Maine homes. Most cotton came to the state in the form of machine-spun yarns. The center of this manufacture was Great Britain, which was using machinery invented particularly for this process in the 1760s. Maine homes contributed labor, not raw materials.

Cotton spinning technology was perfected in New England through the work of Samuel Slater and his followers, who created cotton yarn spinning mills on the English model beginning in the early 1790s. Cotton yarns from England were already flooding New England and the American cotton yarn mills virtually ended the use of flax. The bottleneck in the cotton process was not making yarns, but weaving cloth—a process first mechanized some three decades after yarns were being spun in Rhode Island. In this period, the home weaving of cloth was absolutely required and home labor was a critical factor. In 1810 some 800,000 yards of cotton cloth were produced in Maine, nearly all of it using machine-made yarns woven in family homes.

At first, machine processes were incomplete and the role of domestic industry was actually increased. The first decades of the nineteenth century saw the greatest increase in the use of household labor. This was the most productive period for the hand weaver and all of this was due to the influence of cotton. Yarn mills wanted to produce cloth and these mills contracted with household weavers under a piecework arrangement to produce this cloth in their homes. This has been called the "put-out" system because machine-made yarn was "put out" to be woven in homes. Although this was a trade in yarn and cloth, Maine homes actually sold the family's labor. Yarns were consigned through storekeepers in ever-widening circles from the mill and as mills themselves grew in number, so did the circles of distribution. Thus, weaving was promoted even in Maine though it was, for the moment, distant from the centers of cotton yarn spinning. In 1810 it is estimated that only about 2 percent of cloth was made in factories and as late as 1828 *Niles Register* commented that "low priced cloths…are not even imported to any great extent; they are not even made in factories, but are the products of household manufacture."

The manufacture of cloth was an important aspect of Maine home life and one that touched most households. While the gross production of the state's textiles production remained relatively modest, the value of home cloth production between 1820 and 1830 exceeded $1.5 million. This amount was equal to 40 percent of the value of Maine's manufactured products. (Shipbuilding ranked second.)

Advances in technology inexorably reduced reliance on homes and home weaving, and many tasks were moved to the factory starting around 1820. Power looms were destined to revolutionize the making of cloth and change the nature of work in the home. In addition to mass-produced garment making, the relative abundance of cloth had two important results: (a) transfer of many weaving operations to the factory transformed the work of the itinerant professional weaver; and (b) an abundance of rag pieces and scraps fostered new domestic non-woven handicrafts, including embroidery products, quilts and rugs. Among the woven products, the most commonly found today are overshot coverlets, which created bold patterns by weaving woolen yarn so that a portion of each filling yarn rode "over" the body of the piece to produce a geometric pattern. Early power looms could seldom compete with this work and it is there that most professional weavers carried forward their trade in the years of mechanical improvement. The production of cotton calico prints fostered a tradition of quilt making.

Through the eighteenth century, scraps of woolen cloth were sometimes joined in geometric patterns to make a larger finished blanket or covering than could be produced from any of the scraps. Larger cotton print sheets were sometimes filled with insulating material to produce "whole piece" quilts. But it was the abundance of calico print goods (a direct result of power weaving) and especially the scraps generated by the making of clothing that led directly to the new tradition of making patchwork quilts. Of only limited commercial importance, these quilts frequently displayed sewing and artistic skills. The domestic manufacture of calico patchwork quilts illustrates well the influence of textile factories on home handicrafts. By 1850 virtually all quilting done in Maine homes used products of the cotton factory. Handicrafts were increasingly defined by esoteric qualities not related to the home as a working environment. In this sense quilts and hooked rugs grew distant from their origins and joined the growing assemblage of the "delicate and beautiful handiwork of females." These grew from the home as a "seat of refinement."

Since weaving cloth using power looms was similar to weaving them at home, it is not surprising that household labor was absorbed by factories. The tasks of servicing both power and hand looms was similar and experience with one was transferable to the other. Meanwhile, though, the tradition of home manufacture, particularly of woolens, and the presence of the raw materials and the labor in the home helped sustain home production throughout the century.

The focus of Maine's history has long featured the work of men. There has been difficulty in counting the products of the home, such as textiles, and this has tended to diminish appreciation of the role of women. Moses Greenleaf alluded to this in 1829 when he wrote that the manufacture of cloth was an "important manufacture…conducted chiefly in private families; and it is well known that it is confined almost wholly to the female part of the families, to whom other modes of profitable employment are generally not open; and a large part of whom, without this manufacture, would probably have the opportunity to contribute but little to the General wealth of the State."

In considering the home as a workplace, we should include activities of substantial economic value, and also those of aesthetic importance that enriched wardrobes and contributed much style and color to the home. This is a different tradition, which includes samplers, stitched collars, embroidery, crewel embroidery, upholstery and bed hangings, quilts, coverlets, bonnets and numerous items of interior and personal adornment. Unlike the generic products of the loom, most of these items were personal statements of both skill and creativity. Pride of authorship often distinguished these products and sometimes led to their exposition at craft fairs.

Shown here is a rural hand-weaver in the St. John Valley, circa 1880. *Courtesy Maine State Museum.*

The weaving of cloth was often an example of boring drudgery, and the more affluent families sometimes deferred this work to poorer neighbors. But because of the dominant role of women this work has often been grouped together with creative pastime work, also dominated by women. This includes confusion with many leisure-time hobbies, including needlework, quilting and rug hooking, which actually differs markedly from cloth production. Many of these tasks flowed from the instinct to foster the graces of young women.

Both the continuation of "domestic handicrafts" and the emerging commercial importance of the home can be traced back to the eighteenth-century tradition of making samplers. American samplers, and those found in eighteenth- and early nineteenth-century Maine, were the end products of a tradition that stretched back to the 1500s. In New England, generally, the survival of this handcraft form has been attributed to a number of objectives: training of young girls in needlework skills; cataloguing stitch patterns for later reference; education and moral training through the copying of scriptural verses; and showcasing needlework skills impressive to the parents of potential suitors.

The earliest surviving samplers from Maine date from the third quarter of the eighteenth century and, at this point, we find the samplers of Charlotte Illsley of Portland and Lucy Cushing of Turner. Alphabets were usual motifs and genealogical records are often found as well. Mary Jane Barker, a neighbor and childhood friend of Henry Wadsworth Longfellow, produced a sampler that included the names and dates of children of Thomas and Sarah Barker (Portland, 1818), and Elizabeth Hawkes (Windham, 1815) included the births, deaths and marriages of the Hawkes family.

The average age of girls making samplers was eleven, but some surviving examples represent the work of girls as young as seven (Eliza Goodwin, South Berwick, 1801). Mostly samplers have survived in families showing economic security and social position, but in addition to the influence of families, many were produced as a byproduct of boarding schools and female academies. So we find samplers like that of Eunice Purington (East Harpswell, 1810), whose work was "Certified by Temperance P. Jackson, Instructor."

The story of domestic handicrafts is notably short on statistics, but pride of authorship often distinguished these from the generic products of the spinning wheel and the loom. Needlework skills were ubiquitous in eighteenth- and early nineteenth-century societies. Cloth and clothing were relatively expensive, treasured and expected to be long lasting. In this world embellishments were expected and even lavish embellishments were deemed appropriate. In the eighteenth century "more" was "more." Embroidery stitches on vests and pillowcases, crewel embroidery on bed hangings and rug work were not deemed wasteful. In a period when the way one dressed was a mark of social rank and aesthetics, labor spent on clothing at the hearth must have seemed worth every agonizing stitch. The eighteenth century appreciated excess.

Needlework activity in Maine homes had more social than economic implications. The rich and poor alike invested their time to display these feminine skills—hiring others would have missed the very point of the work itself: the demonstration of their "education in the genteel manner." This is illustrated by entries to the "Charitable Mechanic Association" fair in Portland in 1838. This group had been formed the previous year in an effort to give the city's artisans a social standing that could compete with the influence of the merchant classes. The fair, patterned after that of Boston in 1837, was intended to be an exposition of manufactured wares produced by the

This photo from Center Lovell around 1900 is a reminder of the difficulties faced by home weavers before electric lighting. *Courtesy Maine State Museum.*

male membership. Most interesting for us, though, was the extensive display of the "delicate and beautiful handiwork of females." This display far exceeded the number of items submitted by male members of the club. In all, over three hundred examples of "Muslin and Silk Embroidery" were appraised by a team of seventeen judges. In a remarkably egalitarian spirit symbolizing the spirit of the New Republic, the needlework products of the Maine home provided a common ground of understanding and respect that transcended, at least for a brief moment, other considerations of social position and status. Wrought collars, cuffs and capes formed the largest category of items. Handkerchiefs, shawls, scarves and veils were also numerous. A silver medal was presented to Miss Mary E. Hopkins of Portland for a "wrought collar; very superior—best work," while one of twenty-four diplomas went to an unidentified twelve-year-old handicapped "little girl" from the orphan asylum. While entries often reflected attention focused on the self, occasional demonstrations of selflessness transcend status and are compelling.

In addition to status, the skills shown in making embroidered products were an important personal statement in which the quality of the stitching was thought more important than originality. This body of handicraft was an important bond in the community and one characterized by fellowship and sharing. The consistency of design motifs, and the focus on quality stitching, supported the use of paper patterns (purchased or borrowed) that were often exchanged like so many recipes. The reliance on paper patterns was destined to lead to other embroidery spin-offs, such as rug hooking. There was a broad participation in needlecrafts and, where leisure permitted it, the formation of special sewing circles. The 1838 Mechanic's

This quilt with a tulip or Virginia Lily design is thought to have been made by Amanda Kaler of Waldoboro in the late nineteenth century. *Courtesy Maine State Museum.*

Fair included entries such as the Missionary Sewing Circle, the First Parish Sewing Circle, the Anti-Slavery Sewing Circle, the Free Street Sewing Circle, the Park Street Sewing Circle, the Belfast Benevolent Society and the Female Sewing Circle of Bangor. In addition to embroidery itself, textile handicrafts exhibited included rugs, carpets, quilts, counterpanes and "net work." Of these, the largest category beyond "Muslin and Silk Embroidery" was rug work, although most items were not actually rugs or carpets at all. Exhibited here were various examples of embroidery on furniture, screens, stools, "ottomans" and "crickets," as well as fifty "lamp mats." The artwork found here included the usual catalogue of florals, cornucopia, baskets and fruits—similar, indeed, to the motifs common in theorem painting in this period.

Previewing the future, there were some actual floor coverings and the most interesting for us is the group of thirty-five "made of rags." This coming handicraft grew in direct proportion to the presence of inexpensive woolen and cotton "rags" in the community. As early as 1807 Hallowell's Martha Ballard had "over hauled wollen rags; cut some for flocks and some for carpet filling." At this time, these rags were probably scraps of homespun clothing, but by 1830 many of the scraps came from local woolen mills: Miss Whittier's "mat and rug" in 1838 was, for example, composed of the "head ends of Broadcloth," a waste-product of woolen manufacture. The expansion of factory manufactures of both cotton and woolen cloth aided and promoted the expansion of another handicraft tradition still centered in the home. Since textile factories in Maine grew rapidly after 1850, the abundance of rags led to a heyday in rug production even while many traditional home tasks were absorbed by nearby factories.

All the common techniques to make rag rugs—weaving, braiding and hooking—were used in Maine homes. Woven rugs often substituted rags for filling yarn while still using the surviving hand looms and techniques of common "tabby" weaving formerly used to produce counterpanes, coverlets and all "put-out" work from cotton mills. Wool could be used, but most surviving Maine examples employed inexpensive calico print cotton and, in fact, factory-made cotton cloth gave the hand looms a new lease on life in the Maine home.

By comparison, braided rugs were relatively scarce in nineteenth-century Maine. Secondhand clothing was not the stuff of this handicraft. Instead, rug hooking relied on "remnants" more than rags. The use of relatively uniform strips of cloth allowed rug hooking to follow the same patterns as embroidery and also use paper or stenciled patterns. As early as 1738 the *Boston News Letter* advertised the availability of "all sorts of beautiful figures on Canvas, for Tent Stick [*sic*]." It followed logically that hooked rug patterns could be copied and the availability of burlap permitted development of the modern hooked rug technique and fostered distribution of the rug "stenciled" products. Using brass stencils, it was possible to market pre-stamped burlap marked with designs ready for finishing with woolen remnants.

The contribution of one Biddeford entrepreneur, Edward Sands Frost, revolutionized rug hooking in Maine. Frost began the pre-printing of burlap as early as the 1860s. Starting in the immediate neighborhood, Frost developed an inventory of over 180 patterns. Writing in 1888, Frost recalled that "news of my invention of stamped rugs, spread like magic, [and] I at once became Frost, the rug man." There was a pent-up demand for stamped burlap patterns. Since Frost bought hooked rugs and copied them as stencils, his work helped perpetuate hooked rug motifs and make Frost patterns pervasive in the rug manufactures of Maine during the last third of the nineteenth century. Hooked rugs totally overshadowed shirred and yarn-woven rugs as burlap patterns and abundant woolen remnants promoted a great period of rug hooking stretching into the twentieth century. This handicraft was promoted by the spread of woolen factories, as quilting was similarly encouraged by the rise of the cotton factory.

During the eighteenth and early part of the nineteenth centuries, quilting was commonly composed of outer layers of coarsely woven woolen fabric with a stitched-in filling of loose woolen fibers. But in addition to the "white-on-white" tradition, there were also quilts that featured a patchwork design of smaller shapes and colors. Some eighteenth-century survivors show the use of square or rectangular woolen scraps. In 1785 Moses Davis of Edgecomb noted that he "had a number of Women to Quiln [*sic*] a bed quilt," which was probably made of rectangular woolen patches.

This Waldoboro rug was hooked by Sarah Bagley as a wedding present in October of 1914. *Courtesy Maine State Museum.*

This photograph, taken in or near Androscoggin County around 1875, shows a typical mid-century Maine farmhouse. *Courtesy Maine Historic Preservation Commission.*

It was through the influence of calico cotton cloth, however, that the traditions of patchwork and the patterned quilt stitchery were united. The result was the patchwork quilt, in which small pieces of calico cloth were appliquéd to form a wide variety of mostly geometric patterns. The patchwork quilt, like the rag rug, became popular in direct proportion to the increased abundance of the raw materials from textile mills. In the case of patchwork quilts, the raw material was calico printed cotton, and this nineteenth-century phenomenon was based on machine-woven cloth. These prints were not commonly made in America until the 1820s and by 1838 the Mechanic's Fair exhibited some twelve calico quilts. Some of these were likely not patchwork, but that submitted by Miss Eleanor A. Cole was "made of small pieces" and this was a novelty worth mentioning. By the fair of 1855, however, all nineteen quilts were patchwork.

The rise and growth of both cotton and woolen factories replaced home manufactures one by one. Hand spinning was the first to go, the weaving of lightweight cottons was next and so on throughout the century. Home manufactures adjusted continually to this changing reality. The use of the home as a workplace was well entrenched and, accordingly, hooked rugs and calico quilts arose in the empty space created when factories replaced some home tasks. A result was the rise of a handicraft tradition born less of necessity than of leisure, as the home became a "seat of refinement"; as "fancy work" replaced utility; and as beaded shoes, shell baskets and wax fruit took center stage.

There was a long-standing disparity separating the production of cloth from the various handicrafts, and this may have contributed to the virtual exclusion of many domestic products from the Mechanic's Fair. Products of home looms were notably missing from the fairs of 1838, 1855 and 1860. Coverlets were notably absent; of nearly one thousand entries during those three years, there was only one "coverlit." This was in the golden age of coverlet weaving during this age of homespun. But by 1838, most high-quality coverlets were probably made by professionals and lesser products were much too practical to be handicraft entries.

Inexorably, as the standard of living increased in Maine, domestic drudgery was preempted by the work of the factory and professional weavers produced an increasing number of handicrafts that were associated with leisure, not work. This shift has helped obscure the home as a true workplace and hide the role of women within viable home economies. We must look beyond the Victorian mountain of moss castles, beaded shoes and shell pyramids to a world that lies just beyond; a world no better illustrated than through the manufacture of clothing. While handicraft activities may not have sustained the home as a workplace, the production of clothing certainly did.

Clothing

The impact of machine technology most commonly preceded reorganization of the work and mass production efforts. For example, use of the power loom led to cheaper and more plentiful textiles of many kinds, and it helped produce larger wardrobes. Conversely, though, the case of mass-produced clothing came before the distribution of the sewing machine. By the time these machines appeared in the 1850s, the pattern of sewing in the home was already very well established.

Unless home-woven cloth was intended for sale, this was seldom the final objective. Some cloth was, indeed, intended for bedding, upholstery and draperies, but mostly it was probably used to sew clothing. This, too, was often a task conducted in homes. In the eighteenth and the early part of the nineteenth centuries, wardrobes were small, chests-of-drawers were expensive and most houses were not constructed with closets. Clothing was often ranked high in the value of most estates. Status and relative wealth were important and clothing was a portable way to make this clear to everyone. Wardrobes tended to be elegant, refined and meant to last. Aside from houses, barns and sometimes livestock, clothing said most about status and wealth. In this agenda, quality was celebrated.

Clothing was the most widespread of Maine's home manufactures during the eighteenth and nineteenth centuries. Beginning as a traditional household activity to serve family needs, the production of clothing grew to become one of the most important economic activities of the state, and certainly the most important manufacture ever centered in the home. Domestic production often included everyday and work clothing, frocks, aprons, shirts, trousers, underwear and knit hosiery. As for more formal attire, professional tailors were never far away, serving the desires of the "powdered wig" set.

Up to the sewing of ready-made clothes in 1830–50, virtually all clothing was custom-made. The yardage of cloth needed was commonly purchased from a dry goods merchant. At home, or in the hands of local tailors and their helpers, particular items of clothing were made from this material on a custom basis, or older items were remade to fit the changing family needs.

The home production and repair of clothing was well recognized, and there was significant trade in the repair and resale of clothing. But it was in the manufacture of mass-produced, ready-to-wear garments that Maine homes—and Maine women—played a leading regional, even national, role. The discovery and exploitation of the labor resources in the home were exploited by interests beyond the state. Production of women's clothing was more specialized and this work remained a custom product for another half-century while the manufacture of men's garments, particularly those of mariners, was increasingly mass-produced. Growth in the clothing industry, especially between 1830 and 1890, was fostered by availability of machine-made clothing fabrics and, later, by the invention of the sewing machine. An abundance of cloth led to larger wardrobes, greater attention to current fashions and a reduced cost per garment. Sewing machines later accelerated this development and helped sustain international fashion trends reaching from Europe to local store shelves. Productivity was greatly increased by use of machines, but the process of making ready-to-wear garments had been established at least twenty years before.

In the beginning, the improvement of machinery did not do much to change the organization and distribution of many consumer products. But in the case of shoe and garment production, improvements of machines followed a reorganization of these manufactures. In the case of cloth production, power weaving led the way toward less expensive and abundant consumer products. But in the case of standardized garments, the emergence of mass-produced, ready-to-wear markets emerged in the period from 1830 to the 1850s without significant promotion from the sewing machine.

Up to the first decade of Maine statehood, ready-made clothing was largely unknown and limited to some low-cost coats made for use by seamen. Most all clothing was custom-made, expensive and more practical and long lasting than ephemeral and fashion-conscious. Tailoring was an important part of producing any garment in the home, and since sewing skills varied this work was often given over to local skilled tailors and "tailoresses."

Except for embroidery embellishments and splashy handicrafts, it was easy to take clothing for granted. The stitches needed to make even an ordinary man's shirt were largely unnoticed, thankless work leading to products worn in the fields and not judged at any mechanic's fair. Drawing attention to this work, and arguing in favor of the new sewing machines, *Harper's Bazaar* in 1869 counted the stitches needed to sew a shirt:

> *Stitching the collar, four rows, 3,000; sewing the ends, 500; buttonholes and sewing on buttons, 150; sewing the collar and gathering the neck, 1204; stitching wristbands, 1228; sewing the ends, 68; buttonholes, 48; hemming the slits, 264; gathering the sleeves, 840; setting on wristbands, 1468; stitching on shoulder straps, three rows each, 1880; hemming the busom, 393; sewing the sleeves, 2335; setting the sleeves and gussets, 3050; tapping the sleeves, 1326; sewing the seams, 848; setting the gussets in, 424; hemming the bottom, 1104. Total number of stitches, 20,530.*

The nineteenth century brought important changes in clothing manufacture that were organizational, technological and stylistic. Under the influence of the factory, the price of yard goods declined relative to family income and with this decline came a democratization of clothing styles. By bringing more fabrics within the reach of the average citizen, the elite connotations associated with expensive clothing slowly disappeared in the styling of men's clothing. Long pants had, in the eighteenth century, been associated with the working wear of seamen and laborers, while "gentlemen's" clothing clung to the tradition of knee britches and knit hosiery. But by the time of Maine statehood, all ranks in the society were fitted with long woolen trousers using products of the woolen factory. The dilution, or extinction, of social aspects of men's clothing reflected underlying American values in the age of Jefferson and Jackson. Once this hurdle had been passed, the way was open for the rapid development of merchant-tailor shops whose products were increasingly more standardized than custom, and whose markets expanded to the limits of the neighborhood to the world of regional and even national commerce.

Factory-like organization diminished the importance of apprenticeship in the traditional sense. A tailor, and possibly one or more male assistants, was commonly joined by a larger number of women. As late as 1850 there were still some 587 tailors in the state, but many times more women were also employed by these shops. Messrs. Hyde and Crawford of Calais, for example, had 10 women at work, and Peter Cornell of Damariscotta employed another 10 in sewing some one thousand garments. If not for the influence of larger, regional merchants and distributors, this work might be listed among the state's artisan workshops.

The 1838 Mechanic's Fair caught the beginning of the ready-made clothing business in Maine. William C. Beckett, one of the first "off the shelves" manufacturers in the state, won a silver medal for "a Coat, of elegant finish—the sewing first rate, the collar light and useful," as well as companion pantaloons and vest, exhibiting under "Gentlemen's Wearing Apparel." While most women's wear remained custom-made, the mass production of men's suiting was destined to define ready-to-wear clothing in the decades ahead. By isolating garment assembly from the cutting and distribution of standard garment parts, stitching remained a home activity most often conducted on a piecework basis. As the put-out weaving of cotton goods came to an end, the enormous labor reserves of these home workers were available. Power looms increasingly relocated cloth production to the cotton factory after 1830, and many young women moved to these factories to run these machines. But Maine households now also provided labor in support of a new put-out system: ready-to-wear men's clothing.

This new "sale work" began in the 1830s and saw a major period of growth in the 1850s and 1860s, and the sewing machine was not initially a major force leading to this. Some factors militated against this. There was, of course, a long tradition of family work that tended to be subordinated to this practice. Fashion tended also to follow hand-sewing styles. Collars and cuffs, for instance, commonly employed a series of smaller tucks better suited to the hand tradition than to the machine age, as did the remaking of older clothing. But mass production of some clothing, particularly that of men, was fostered by the repair of clothing. Ready-to-wear fashion proceeded from the simple to the more complex: from seamen's goods at the start and finer women's dress goods at the end. Although Elias Howe had patented a sewing machine in 1846, and inventions followed in the 1850s, production in Maine relied on hand sewing. By the time sewing machines became common during the 1850s, the pattern of distribution and sewing in Maine homes had already been instituted.

Garment manufacturers were tailors or merchant-tailors like Charles Hilton of Winthrop, who produced 206 sets of pantaloons, vests and coats using six women in 1850. At the same time, James Merrill of Sullivan employed seven women to make "144 thick cloth coats" and "300 thinner cloth coats." In New Sharon, tailor James Diggles used twelve women to produce some 2,000 coats. Facing large quantities and variable quality, many traditional tailors found fault with these products, noting that a ready-to-wear suit could be identified from many paces. In the egalitarian world of the New Republic there was great demand for ready-made clothing and shoes. In both cases, the labor of the Maine homes was ready to answer.

Managers of Maine's ready-to-wear men's clothing were essentially middlemen who secured patterns and cut clothing parts from Boston area suppliers and paid local women for sewing. Cloth was provided by textile factories and they were cut, sometimes several layers at a time using clothier's shears.

In the Augusta-Gardiner area, W.A. Lawrence advertised in July of 1852 that "50 Coat Makers Can Find Employment of Boston Work," and the following March he renewed his call for "100 Coat Makers." A.A. Luce offered "the highest wages" and advertised in 1859 that he sought "3 to 500 Coat Makers, to whom the best prices will be paid, and constant employment given." By 1860 Luce was employing 4 men (probably teamsters) and 800 women to produce "clothing for other houses." In Hallowell, Samuel W. Huntington maintained nine horse teams to pick up and deliver garment parts in the local area.

In Buxton, Samuel Hanson was the undisputed leader of the garment business and through this "Boston Work" he became rich, as this enterprise (30,000 coats a year in 1850) was recognized to be the town's principal business with a total employment of 1,200 in the women's ledger. Sewing machines were not yet important. But by 1860, the largest merchant-tailor in Maine was Josiah A. Bucknam of Minot, who claimed to employ more than 1,500 women.

During the 1850s, and particularly the early 1860s, sewing machines impacted this enterprise; of all nineteenth-century inventions, few had greater impact than the sewing machine. Emerging after many decades of experimentation, the sewing machine was a masterpiece of engineering that increased worker productivity by a large measure and endured with little change until the present day. As early as 1860, Isaac Brackett of Sanford equipped all three hundred women in his employ with sewing machines, even though home sewing remained important in many rural areas; Nathaniel Pendexter employed three hundred women in Parsonsfield but only supplied ten sewing machines, while his brother, George, employed one thousand women with only twenty sewing machines. Sewing began at this time to be centralized in workshop environments, but old ways died hard and the role of families working at home was long established through the 1840s. Although Elias Howe's sewing machine was not patented until 1846, his effort was but one of many and the 1850s saw many versions of lock-stitch and chain-stitch machines emerge, as these increased worker productivity by ten to twenty times. Ultimately these machines revolutionized the sewing of clothing. The sewing machine was among the best-engineered and most successful machines ever developed, but unlike the loom, there was no way to gang these products: clothing was still sewn one piece at a time, by one worker at a time. In the case of sewing, more than any other activity, the hourly cost determined the location of this work.

During the first decades of Maine statehood, ready-to-wear clothing was virtually unknown. Some trade emerged in the semi-tailoring skills involved in darning and mending clothing as well

Shown here is a long-abandoned house that belonged to the author's ancestors in the Deering neighborhood section of Springvale. Originally built by the Butler family around 1770, houses like this were common farmhouses. Seldom idle, family members often used houses like this as workplaces. *Courtesy Gerard Rivard.*

as remaking "hand-me-downs." Still the products were inconsistent and the presence of nearby tailors was an important asset to most households. The home manufacture of everyday clothes, work clothing and undergarments were common tasks in most households. Among the items frequently found enumerated in inventories are petticoats, nightclothes, frocks, aprons, shirts, trousers and knit hosiery. Producing formal attire was not commonly a domestic activity in the eighteenth century and stylish clothing was provided even in remote rural areas.

Tailoring skills were customarily learned through apprenticeships with local tailors and, thus, young women were trained in the making of finer clothes. Accordingly, Martha Ballard of Hallowell noted in 1792 that her daughter Dolly had "returned from her apprentiship with Mr. Densmore [a local tailor]." Not surprisingly, Dolly was engaged later to make " a waist coat for [her brother] Cyrus."

The importance of family labor inexorably declined under the influence of machine technology and the factories it spawned. But in terms of the Maine household, the most significant of these machines was, in fact, the sewing machine. While the production of these machines was insignificant when Isaac Singer patented the first practical machine in 1851, by 1860 the production of sewing machines nearly equaled the value of all the textile machines destined for factories, and the use of these machines became general. This was truly a revolutionary technological advance, marketed to women, made throughout America (including Maine) and a new tool suitable for both the workshop and the home. This machine did not displace the domestic worker, but by 1880 hand sewing had largely become a thing of the past.

The objective of virtually all machine improvements was greater speed and increased productivity. Better water wheels and better engineering allowed machines to run faster and make more cloth while the use of "stop motions" permitted supervision of multiple machines that were each smart enough to shut themselves off if something went wrong. The sewing machine, too, increased worker productivity, but there was a definite limit that sustained the importance of the individual sewers, whether at home or in workshops. Making a coat, pantaloons or a shirt was still achieved one at a time. Though the stitches might come faster, the tasks were still individual and sewing remained more labor-intensive than making the cloth itself. The production of cloth soon outdistanced the sewing of clothing and a reasonable machine speed began for cheaper labor. Outsourcing emerged during the second half of the twentieth century.

The Survival of Domestic Manufactures

No sooner had spinning and weaving ceased to be common needs of communities than this work became a romantic memory of simpler times. Drudgery was forgotten and replaced by nostalgia for a past that had never really existed. Hand spinning, weaving and even sewing were lumped together with the world of sewing circles, hobbies and creative pastimes, while the surviving hand tools of these domestic activities were assembled by local museums, historical societies and community history pageants.

II.

The Mechanic's Workshop

In the "good old days" of which we hear so much about, the chief aim in manufacturing was to obtain durability. Now this state of things has all changed and we are not satisfied with articles which will endure simply but we must have showy and attractive as well. This change of taste has made employment for numerous men and opened up new trades and branches of business.

Somerset Reporter, May 22, 1879

The commercial exploitation of labor in Maine homes tended to make the activities of domestic manufacture somewhat specialized. In this regard, the home environment sometimes resembled the artisan's workshop, but the tools of most craft manufactures were usually assembled elsewhere, by convention or because they created noise and mess. The artisan's workshop reflected a commitment to specialization that often distinguished artisan products from those of the home. At first, craftsmen were considered more important than either tools or machinery. In the eighteenth and early part of the nineteenth centuries, workshop environments presupposed the leadership of crafts people.

In the eighteenth century, the density of Maine's population was seldom a sufficient motive to establish full-time workshops. More than 80 percent of Mainers called themselves farmers, but products of the soil were rarely adequate to provide a family's entire income. This helped establish a hybrid source of work and income featuring workers whom Tench Cox referred to as "farmer-artisans." Commerce, not local craft manufactures, supplied most needs.

There existed a strong tradition of part-time activities integrated with the farmer's work year and, at the same time, there was often insufficient market demand to support full-time craftsmen. This was true of Moses Davis of Edgecomb, both a farmer and a cabinetmaker. In addition to agricultural duties, Davis reported in his diary for September 11, 1783, that he "workt in Shop on Capt. Dunton chest." At the same time, Davis worked as a farmer, made a "great plow," finished a chest for John Place, built a bedstead, made a "standing stool" for a Mrs. Decket and helped build a kitchen in Wiscasset.

Moses Davis's life epitomizes the romantic notion of the Yankee jack-of-all-trades, a colonial craftsman. This is an appealing view but, alas, it has little to do with the story of Maine's mechanics and workshops. More often than not, the craft work of farmers helped produce staples such as shingles, clapboards and barrel staves. Meanwhile, many artisan workshops resulted from regional, even international, commerce.

Isolation was sometimes a motive for local manufactures in America. Inland communities, lacking water-borne transportation, tended to protect small-scale craft shops from larger, distant competitors. This was not generally the case in Maine where the sea and inland waterways helped the distribution of raw materials and finished products. Never far from these major lanes of commerce, Maine enjoyed an advantage in fostering commercial development. At the same time, this commerce also diminished the importance of artisan manufactures and, as Moses Greenleaf noted in 1829, Maine was "yet in its infancy" regarding manufactured products. Statistics supported this conclusion. Although there were notable gaps in accounting for "manufactures devoted solely to the supply of the population in the immediate vicinity," there were only 1,768 mechanic workshops in Maine in 1820 and this amounted to just one workshop for every thirty-three families.

The products of the workshop have occupied a favored place among the manufactures of the society, while products of the home, as well as mills and furnaces, were more often generic. Many shop products exhibited enduring regional, community and even individual characteristics. With little exception, these products of the workshop provide most antique collectibles. This fact, in turn, reinforces the myth of the "colonial craftsman" and Moses Davis–type figures.

The custom nature of some manufactures, bespoke production, kept business organizations relatively simple. Blacksmiths, shoemakers and furniture makers often worked as jobbers, producing items on order. This service to a local audience helped maintain local, regional and even individual peculiarities. There was limited investment and little risk in this work, but also limited income security and little potential for expansion. Bespoke manufactures remained important in the shops of tailors, seamstresses, blacksmiths, machinists, wheelwrights and, in a modern age, service stations. Much bespoke manufacture was extended by work "immediately connected with and indispensably connected to, the ordinary pursuits of agriculture."

Many, if not most, workshop crafts evolved quickly into the wider production of sale work. Here, commodities were produced at the pace of men and machines and then distributed as widely as possible. This effort was fostered by a preexisting network of inland water-borne commerce and the growing national trade. Although some workshop industries in Maine prospered as a direct result of abundant raw materials, particularly wood, other trades flourished through the importation of basic materials. Thus, both the J.T. Winslow Company of Portland and the Bangor Stoneware Company of Bangor produced pottery in Maine for export beyond the state, but did so using clay shipped by schooner from New Jersey.

In the Service of Agriculture

In 1850 there were some 83,893 oxen working on Maine farms and hauling logs in the woods during winter. In addition, there were over 41,000 horses. Together, these herds offered millions of hooves to be shod. Since there was a reasonable limit to the distance an ox could walk for a new shoe, blacksmith shops remained both widespread and parochial. As Abraham Newbegin of Bridgeton noted, these blacksmiths performed "custom work done in the country." Not including those employed in ship work, 2,700 blacksmiths were involved.

It was, no doubt, this widespread, small and thoroughly local service of the village blacksmith that Moses Greenleaf referred to in 1829 when he wrote:

> *In those branches of manufacture and handicraft trades, which are immediately connected with and indispensably necessary to, the ordinary pursuits of agriculture, and are necessary to be obtained in the neighborhood of every farming establishment, Maine has already made a progress, in general nearly sufficient for its wants, and perhaps in most cases quite so.*

Blacksmithing

There was little business evolution within service-oriented blacksmith shops. Custom work and rural orientation tended to hold the blacksmith apart from the world of regional commerce and preserved, instead, an archaic level of craft development. Although the care of animals did most to preserve the dispersion of blacksmithing, the custom work and associated tool-making and carriage work generated the bulk of the income. A typical one-man shop was estimated to generate between $500 and $700 per year, and the most frequent activities reported were horse and ox shoeing and "wheel tiring." Elias Wate of Wilton, for example, worked with a helper in 1850 to produce three hundred pairs of horseshoes worth $75 and one hundred "setts of ox shoes" worth another $75. But other articles of custom work, probably including the making of wagon axles and edged tools, amounted to $700.

The village blacksmith shop was not, however, an accurate symbol of the nineteenth-century craftsman's workshop. Most manufactures participated in an evolution that was sometimes technical, but always related to changing patterns of labor organization and craft specialization. Shops to manufacture plows, agricultural implements, carriages, edged tools and machines all traced their ancestry to the village blacksmith shop and evolution came first in the manufacture of plows.

Plow Production

Up to the 1840s, most of the work of making plows was woodwork and so it is not surprising that the "ploughmaker" was most often a woodworker. Ironwork was tributary, but it is here that blacksmiths first participated in the manufacture of ready-made commodities. Even when most forged iron parts had been replaced by cast iron, the plow handles and the beams were often made of wood.

Although Philip Torrey of North Yarmouth, an active blacksmith in the 1820s, was recalled as being a manufacturer of "wooden ploughs," a more typical arrangement was that of Daniel Woodward and John Stone of Gardiner. Woodward, who made the wooden parts, was a "ploughmaker," and he received the needed iron forgings from Stone, who mass-produced them using a triphammer in his shop. Meanwhile in Paris, Maine, another woodworker, Richard Morton, made plows in concert with ironworkers Simeon Hersey and Perez Record.

In 1898 Edward Page and Ernest Mills posed for this picture while mowing in Belgrade. *Courtesy Maine State Museum.*

The production of plows exhibited a pattern of workshop manufacture noticed in all commodities intended to serve the agricultural community—products were distributed from relatively small shops employing only a few men. Timothy Shaw of Exeter, for instance, worked with one helper in 1859 to assemble 190 finished plows and Hiram Doe of Vassalboro employed four men to make 250 plows and 30 cultivators. Working in small shops like this, the manufacture of plows, and indeed agricultural implements, was decentralized and tributary to the needs of the local farming community. The work of four plow makers in Worcester, Massachusetts, exerted an overriding influence on the construction of Maine plows. Exhibiting at Portland's Mechanics Fair in 1838, entries by David Prouty & Co. and Ruggles, Nourse & Mason revolutionized the manufacture of plows.

The impact of Ruggles, Nourse & Mason's Eagle plow cannot be overestimated and in 1855 the Portland judges had to say, "We think no plow can be produced, that will turn a better furrow." Most cast-iron plow manufacturers were licensed under the patents of these manufacturers or made only superficial changes in the basic design. Standardization reinforced the distribution of similar castings—as the foundry in Belfast that offered castings on the Doe pattern.

In the period immediately following the Civil War, the largest plow maker in Maine was the Augustine Hussey Manufacturing Company of North Berwick. Using iron recast from scrapped war armaments (literally beating swords into plowshares), Hussey produced some 800 plows in 1870, in addition to 500 "plow handles." In that same year, over 32,000 plows were made in Maine. Most were made by Maine's thirty-two manufacturers of agricultural implements.

Agricultural Implements

While the manufacture of plows was diversified, the influence of commerce was ever-present. Another category of manufacture arose under the needs of farming—products including a wide range of tools and machines used by the farming community. While artisan workshops of the eighteenth and nineteenth centuries were often characterized by their use of a single material (i.e., iron, tin, silver, pewter), the nineteenth century ushered in the manufacturing unit organized to serve a particular market.

While the manufacture of plows had involved the labor of the woodworker and the blacksmith, later the iron founder, the production of a larger selection of machines required a greater range of trades. The precision of the machinist and the embellishment of the paint decorator were now essential. The blacksmith was indeed versatile, but his was inherently an imprecise craft as compared, for instance, with the clockmaker and machinist.

Threshing Machines

Although the first American patents for threshing machines were awarded in the eighteenth century, it was not until the 1830s that a concerted effort was made to improve these machines. Up to that time, the age-old process of handling wheat included use of the flail and the winnowing basket. Coincidentally, mechanical improvements for this process (and many other agricultural processes) paralleled improvements in cast-iron plows. Maine builders of agricultural implements joined in a national stampede to patent, improve and build threshing machines.

The machine patented by Samuel Lane of Hallowell in 1830 illustrates a common pattern. Lane's machine claimed "certain improvements in the Machine for Thrashing and Cleaning Wheat and Other Grains," which included some modest changes in a preexisting model. Lane was able to sell patent rights for this thresher for $350. The changes were small and, of course, there was a lot of cheating.

Many makers of threshing machines were clustered in the central Kennebec Valley. In 1830 Holmes and Robbins, the pioneer machine shop of Gardiner, began to make these machines for the agricultural implement maker Reuben B. Dunn and, in 1834, a Winthrop manufacturer began to build what was claimed as "the first practical grain thresher and separator ever made." Most builders of threshing machines also offered treadmills, called "horsepowers," which allowed draft animals to provide power for the threshing machines themselves. So it is not surprising that Samuel Lane patented his "endless chain for railway horsepower." Despite claims to the contrary, most horsepowers and threshing machines were very similar—market competition centered on price.

Despite decentralization, Maine sustained its share of the agricultural machine business during the nineteenth century. By mid-century, the shops of Benjamin Davis and Luther Whitman were among forty-nine agricultural machinery makers in the state. Although Luther Whitman employed as many as 70 men in 1889, between 1850 and 1880 the focus of this manufacture shifted west to Cincinnati, Ohio, and Racine, Wisconsin, and as early as 1850 only 325 people were engaged in this business statewide.

Edged Tools

Maine's contribution to the national market for agricultural tools largely centered not on machines, but hand tools, particularly shovels, hoes, rakes and especially scythes. The abundance of lumber resources and waterpower sites sustained this industry in Maine. In 1869 there were no fewer than thirteen small shops producing shovel handles, with one shop in Monmouth providing some forty dozen shovel handles per day. In Waterville, Henry R. Butterfield employed fifteen men to make over four hundred thousand shovel handles each year. While some shops specialized in making rakes, in the town of Dennysville some fifty dozen butter churns were produced each week.

The production of shovel handles, churns and rakes can be traced to timber supplies, but the manufacture of scythes depended on the enterprise of Maine merchants, and for a time in the 1860s the state boasted the largest scythe and edged tool manufactory in New England. Scythes used both metal and wood parts and even when steel was substituted for iron, woodworkers made the handles and cradles. This industry grew from the workshop to the factory stage at the hands of Reuben B. Dunn who, in 1850, was known as a "grain cradle manufacturer." Moving from North Wayne to Oakland, the Dunn Edged Tool Company made fifteen thousand scythes and ten thousand axes by 1864.

With the shift from wood to steel, this industry was obviously misplaced in Maine. The edged tool business was destined to be tributary to steel mills, and in 1889 the American Axe and Edged Tools Company was formed in Pittsburgh with a workforce soon approaching ten thousand

When agriculture ruled the day in many towns, most residents chose to call themselves farmers. It is not surprising that the production of agricultural implements was widespread. *Courtesy Maine State Museum.*

men. The Dunn Edged Tool Company was soon just number sixteen in the growing legion of Pittsburgh plants.

Forest Products

In 1850 most of Maine's working oxen lived and worked on Maine farms, and spent winters drawing sledges of logs to the rivers and streams. The woods, not the farms alone, helped sustain the relatively large population of draft animals in the state. The tendency of all industries to group the manufacture of supporting tools and materials explains the operation of workshops that served the needs of the logging and lumbering industry, or utilized its products. The production of axes is a case in point. This was a major industry in Union where 10,000 axes were made by the Dunn Edged Tool Company, while the nearby Oakland company, Emerson and Stevens, was producing some 36,000 axes of all shapes and sizes at mid-century.

The logging industry fostered the marketing of other tools, including sledges, pike poles and cant dogs. These were perfected in 1858 by Joseph Peavey of Stillwater and remain the standard tool used in the Maine woods in modern times.

The presence of abundant timber resources was seldom a sufficient motive to area workshops. More often, shops were led forward by other industries. The textile industry, for example, suggested the manufacture of bobbins and spools. Not surprisingly, the Lewiston Bobbin Shop produced some two million bobbins each year, while in 1870 the Frederick Richmond Company of Livermore used one hundred cords of birch to make thread spools. *Scientific American* reported in the 1860s that one small spool shop in Maine provided two-thirds of all the spools used by the entire American textile industry.

Bobbin and spool shops used water-driven machinery to make repetitive products. The makers were not colonial craftsmen. Learning from textile mills, hand skills were replaced by production machinery and machine operatives during the nineteenth century.

Specialization and the improvement of turning machines contributed to increased productivity. While H.P. and L.H. Hussey employed eight men and made 75,000 oars in their hand-powered oar-making shop in New Limerick, the makers of clothespins, toothpicks, shoe pegs and friction matches were increasingly mechanized and factory-like. The makers of bobbins and shoe lasts were intermediate between mill industries and the factory towns downstream. Craft shops displayed factory organization borrowed from the shingle and clapboard mill.

Barrels and Casks

In terms of employment and the value of products, all workshop manufactures using wood were strictly secondary to the production of barrels and barrel parts. Coopering, and the related mill industries dwarfed the production of bobbins, oars and shovel handles to near insignificance.

Except for a few uses, wooden barrels are essentially obsolete today. At first settlement, however, all the most important raw materials from the New World were shipped in barrels—the essential containers of the period. In addition to water itself, pitch, tar, turpentine, potash, fish, dried fruits and meat were all shipped in barrels, as were the products of the West Indies: rum, molasses and sugar. Maine was both a producer of barrel parts and finished barrels, producing barrels in excess of the state's actual need. The state boasted more than its share of coopers and many products arriving on Maine wharves were packed in barrels that had previously been made in the state. Increasing quantities of barrel staves were shipped from Maine in the eighteenth and nineteenth centuries. In 1826, for example, one medium-sized port, Saco, sent out over 65,000 staves. Obviously, these barrels were dismantled before being sent. "Shooks," or barrels that had been shaken, were the largest part of Maine's coopering work during the nineteenth century.

Down to 1820, the handmade barrels, including the splitting and trimming of hardwood staves, were dampened and heated to bend them into place. As early as 1815, machinery in Cincinnati was reportedly designed to cut and dress staves and by the 1830s sawing machines had been developed that allowed two men to cut all the staves for one barrel in three minutes. By 1850 stave mills produced blanks at the rate of 8,000 per day. This advance of machinery tended to take coopering from the workshop stage to the sawmill. Many effects of machinery can be seen by reference to the town of Bridgeton. At mid-century three coopers produced over 2,000 barrels in this town. There were also some twenty-six others (coopers, laborers, farmers) who produced over 28,000 shooks using staves hewn by hand.

Abel Whitney of Gardiner supplied barrels for the export of agricultural commodities shipped to the West Indies and California. Among the products leaving Gardiner were dried and green

The hand sewing of men's hats is depicted in this interior view of the Ayer-Houston Hat Factory in Portland, around 1870. *Courtesy Maine State Museum.*

Workers at the Portland Broom Company pose for this photograph around 1870. Broomcorn was imported to Maine from many places in the Midwest, including Illinois. *Courtesy Maine State Museum.*

apples and, believe it or not, eggs. Many of these products did not require watertight containers, and stave sawing machines were frequently used to produce looser, dry barrels. The biggest use of these was for the export of lime from the central Maine coast.

Containers used for lime were tight, but not watertight. Beginning as early as 1733, lime for plaster and brick mortar grew to be the principal activity of Thomaston, Rockland, Warren, Rockport and Camden, and by 1835 there were estimated to be not less than 750,000 casks of lime shipped from the state. By 1850 the export from Rockland alone exceeded 1 million barrels per year and over five hundred men were kept at work in nearby communities to cut and assemble the needed casks. Predictably, the centers of lime cask manufacture were immediately inland from the Rockland area at places where falls of water fostered the construction of stave mills, such as Union, Liberty and Washington. Dry barrels required less skill and they were therefore less expensive—averaging twelve to fourteen cents per barrel at a time when waterproof hardwood barrels were worth eighty-five cents. As new technology was introduced, the craft of coopering was replaced by machines. This transition increased productivity even while the cooper's role was diminished.

Cabinet and Chair Makers

Workshop industries exhibited a range of sizes, sophistication, mechanization, price and product distribution. The generic nature of some products, like barrels, tended to make price the greatest arbiter between these products. The opposite was true of furniture making. Here the commodities produced varied greatly and were influenced by location, price, quality, style, use and needed skills. The availability of veneer woods and a general decline in the cost of clothing were each of particular importance to Maine cabinetmakers.

Machines for woodworking were introduced throughout the century, and these altered furniture styles while replacing craft skills. Artisan workshops began to be like factories long before the widespread use of machines.

In the eighteenth century, a considerable amount of furniture was made by local carpenters and shipwrights. Not surprisingly, the makers of high-end furniture clustered in seacoast cities where they remained in contact with broader market opportunities and the importation of cabinet woods. In Portland, cabinetmakers like Benjamin Radford and Benjamin Isley were leaders of a group that included George Clark, William Haskell, R. Billings, S. Roberts, Samuel S. Rich and Walter Corey. Using mahogany and rosewood as principal materials, the craft shops here produced secretaries, bureaus, washstands, dressing tables and work tables in the Hepplewhite, Sheraton and, later, Empire styles.

The manufacture of mid-range formal furniture was supported by Maine's waterborne coastal traffic. While shops remained small, each was able to offer a full line of hardwood furniture. The middle-class audience accepted changes in furniture construction placing an increased emphasis on hardwood veneers over cheaper, softer woods. This furniture suited small workshops and reinforced the use of pine and basswood as base materials. Power-driven veneer saws also promoted the use of local woods, including birds-eye maple and birch.

The use of veneers dramatically changed the work of most Maine furniture shops, especially in the period of Empire-style furniture. For instance, while Jonathan Raynes of Lewiston used 35,000 feet of boards in the production of chairs and bureaus in 1850, he also used "2,000 veneers." These were often supplied by specialists such as Henry S. Alexander of Hanover, "Master of Veneers," who at mid-century produced 100,000 feet of maple veneer stock.

From the 1830s onward, a significant part of Maine-made furniture used soft local woods decorated with paint. Sometimes this decoration imitated more expensive woods and veneers and also provided new opportunities for decoration. This resulted in a rich tradition of paint decoration in the state that sometimes eclipsed the work of the cabinetmaker. When Benjamin Clark of Harrison made his bedsteads, bureaus and chairs, one hundred pounds of paint were needed.

Many furniture manufactories operated with a full complement of the latest machinery. B.F. Cunningham of Freedom employed only three men at mid-century but his shop was equipped with a planing machine and turning lathes, while Reuben Flanders's two-man shop in Dexter included "3 circular saws, 1 lathe, 1 planing machine." This evolution of manufacturing technology was perhaps best illustrated by an article in the *Bangor Daily Whig and Courier* on July 7, 1843:

> *These thoughts were suggested yesterday on visiting the furniture manufacturing establishment of Messrs. A. & E. Dole & Co. in this city. Twenty years since we frequently met some of our young*

This teapot was made by Allen Porter of Stevens Plains before 1850. *Courtesy Maine State Museum.*

> *friends at this establishment which at that time stood as high in public estimation as any other in the country. But what a change! In the basement where then a large hand wheel stood to be turned by the apprentice when the lathe was in use, which was seldom and only for small articles, there is now a steam engine consuming about half a cord a week, besides the chips made in the shop, with only occasional and slight attendance from one of the workmen, and then the same room will be put into operation a machine for sawing veneers.*
>
> *There is in operation planing machine designed by Mr. Dole which performs the principal part of the planing required for the whole establishment, from a drawer bottom to a bed post. This machine will plane to a thickness on both sides of 2000 feet of boards a day. There is a circular saw and two upright jig saws that work finely in the most crooked lines desirable, leaving the surface smooth enough for a glue joint. There is a boring machine with which a boy in a few hours practice can bore bedstead rails at great speed. There will shortly be put in operation a tenoning and dove tailing machine and a mitre saw.*

Under the influence of machinery and factory-like organization, Maine's chair manufacturers turned out large quantities of relatively inexpensive Windsor-type chairs and settees. In 1850,

These whale-oil lamps were made by William Webb of Warren around 1839. *Courtesy Maine State Museum.*

John C. Butterfield's two-man New Sharon shop produced 2,600 chairs, which sold for twenty-three cents each. Meanwhile, Maine's largest chair maker, Joseph Batchelder of West Waterville, made 8,000 chairs in his four-person shop and sold them for twenty-five cents each.

Furniture making was unlike the manufacture of plows or edged tools and more like the production of generic bobbins and spools—it was seldom a composite manufacture. Despite the need for specialists in upholstery and paint decoration, the manufacture of furniture involved the skills and machinery of woodworking and the raw materials tended to come from sawmills.

The course of Maine furniture-making paralleled that of the whole country. It was influenced by demand from the bottom up and directly related to textiles and clothing. Wardrobes were small in eighteenth-century America and clothing was a prized and expensive status symbol. In this world, homes offered no closets and bureaus were not needed by most of the population. Competition drove cotton and woolen mills to greater and greater productivity and, under the effect of machine-made textiles, the cost per yard declined. Clothing became affordable and bureaus were now needed by many people. Through the use of veneers and paint, as well as an increasing distribution of small shops producing Empire-era furnishings, less expensive furniture evolved to serve needs driven by less expensive clothing. The result was a democratization of both clothing and furnishings.

These andirons were made by William Holmes Webb of Warren in the 1820s. *Courtesy Maine State Museum.*

Clocks, Pianos and Carriages

Mechanization promoted centralization and this, in turn, often led to the decline in Maine's artisan manufactures. Thus the production of clocks, pianos and carriages was never really industrialized in Maine. These manufactures were destined for extinction as they were overcome by forces outside the state.

The decline was not immediate because each of these manufactures passed through phases including the sale of component parts. Clock movements, piano hammers, soundboards, axles and springs supported local clockmakers, piano builders, wheelwrights and carriage makers, respectively. These components provided key parts supporting many "composite" manufactures. Skills declined in the artisan production of clocks, pianos and carriages as the tasks turned from construction to assembly, repair and sales.

Clocks

Until the early nineteenth century, case clocks were both timepieces and items of domestic furniture and their construction included both movements and cabinetmaking. Movements were themselves the result of artisan training, experience and skill, and the period from 1800 to 1830 saw a number of Maine clockmakers, including Nathaniel Hamlen and Frederick Wingate of Augusta; Paul and Abner Rogers of Berwick; Nicholas and David Blaisdell of Falmouth; Lebbeus Bailey of North Yarmouth; Nathan Adams; Daniel Dole; Enoch Burnham; Jonathan Bemis; Phineas and William Quimby; Abiel and Robert Eastman; and James Carey. All these men were true manufacturers who made clock movements (often of wood, sometime of brass and occasionally iron) as well as the casework.

While Benjamin Swan of Augusta continued making clocks in the workshop tradition through the 1840s, the advance of mass-production techniques were pioneered in Massachusetts and

This small lantern is marked "R. Dunham Patent, Oct 24, 1865." *Courtesy Maine State Museum.*

This piece of stoneware was made in Gardiner by Ballard and Brothers in 1854 or 1855. *Courtesy Maine State Museum.*

particularly Connecticut. As early as 1807 Eli Terry of Waterbury, Connecticut, was engaged to produce some four thousand wooden clock movements in a period of three years. Under the influence of two of Terry's workers, Seth Thomas and Silas Hoadley, Connecticut became an emporium for the production of clock movements.

Clockmaking was never industrialized in Maine, however, and the proliferation of mass-produced movements, and later the cases themselves, ultimately drove Maine clockmakers out of business. By 1850 there were only fifteen clockmakers listed in the federal census of Maine, and most were probably engaged primarily in repair work. At the same time, an influx of imported movements meant that the cabinetmaker became the true manufacturer and clocks were embraced within furniture lines. In Belfast, furniture makers like Daniel Howard and Thomas Town assembled clocks in this last gasp of Maine clockmaking.

Pianos

The fate of clockmaking in Maine was paralleled by piano, melodeon and organ production, except that these manufactures came to Maine relatively late and survived into the 1880s. In 1850 there were three piano makers in Portland: Calvin Edwards, E.E. Kendall and George Clark. Also in Portland, William Hastings and J.D. Cheney made organs and melodeons and there were other shops in Farmington and Hallowell.

Carriages

Long associated with blacksmiths and wheelwrights, carriages and wagons enjoyed a long tradition of local, even rural, manufacture. Carriage and wagon making exhibited a history similar to that of clocks and pianos except that mechanization progressed much more slowly, specialty manufacture of component parts arose within the state and the industry was resistant to centralization until the era of Studebaker and Sears, Roebuck.

In the production of wagons and carts for the farm, blacksmiths and wheelwrights were central, and the strength and durability of wheels determined success. Wheels originally needed the cutting of hubs, the shaping of spokes, the steaming and bending of fellows and the assembly of these parts into a tight, flexible wheel. This was no simple task and, accordingly, wheelwrights were able to sustain a handcraft tradition well into an age otherwise dominated by the machine. Wheelwrights were often producers of farm vehicles. Thus, Gardiner Phillips of Augusta made some twenty wagons in 1850.

The manufacture of "pleasure carriages" was quite another matter. Limited to the ownership and use of a community's wealthy gentlemen, the owners of these vehicles demanded not only serviceability but also style and even opulence. Smartly appointed and decorated coaches were, in their day, symbols of excess, statements of status similar to clothing. Not many pleasure carriages were owned in Maine, and not many were made in the state either. In 1820, for example, the value of all Maine-made vehicles, except for wagons and carts, was only $12,000.

The 1830s saw an increase in the number of pleasure vehicles made in the state. The most popular wheeled vehicle of the period was the chaise, or, literally, the chair. This form, popular in

Pattern makers are seen here at the Portland Stove Foundry around 1890. *Courtesy Maine State Museum.*

cities during the 1770s and 1780s, was built on a base of only two wheels and it stood high above the road surface with cantilevered wood frames and leather belting providing minimal suspension and shock absorption. Although considered obsolete in some American cities as early as 1820, the chaise was well suited to rural road conditions and it continued in use for decades. Three examples of Maine-made chaises were exhibited in Portland in 1838 and the entry of J.J. Russell of Portland was judged "extremely neat, very light, well finished…a superior article."

When Hiram Hatch of North Yarmouth made that town's first "bellows top chaise" in 1827, he probably used the combined skills of the cabinetmaker, upholsterer and decorative painter. Making expensive carriages for the wealthy was similar to making clothing and furniture for their homes. The luxury view of carriages continued through the first quarter of the nineteenth century but ultimately gave way to the democratization in the age of the New Republic. By mid-century, carriages and sleighs were relatively commonplace and affordable. As a consequence, by 1850 there were 382 "coachmakers" in the state in addition to 348 wheelwrights. Between 1850 and 1900 no less than 1,500 called themselves "carriage makers." This proliferation of carriage manufacturers encouraged the specialty manufacture of carriage components. The mass production of wheel hubs, for instance, was located in Westbrook as early as the 1840s when Moody Morse produced ten thousand "Hubbs" per year. Meanwhile, the production of steel springs was underway in the Readfield shop of James Williams, who produced four thousand pounds of springs at mid-century.

Bangor photographer Merritt Jordan produced this ambrotype of a local shoemaker in the late 1850s or early 1860s. *Courtesy Maine State Museum.*

The availability of mass-produced component parts encouraged a bewildering array of "carriage makers" in Maine. One of the largest, Charles Kimball of Norway, employed eighteen men in his water-powered carriage "manufactory" to produce seventy-five carriages and sixty-four sleighs in 1850. There were some centers of production, including Union, where Collins, Wingate and Sittee made three hundred carriages and sleighs in 1860. In 1877 Wingate, Simmons and Company made thirty "top" carriages, six buggies, forty wagons and eighty-three sleighs. The increasing market for vehicles was met by an expansion in the number of artisan workshops producing them, rather than the opening of larger, centralized factories.

Workers in Tin and Pewter

The distribution of tin workers in 1850 (some 468 in the state) indicated only a modest relationship between production and local needs. The productivity of both pewter and tin tradesmen extended beyond local considerations in every case and the sale of the resulting consumer products did not rely on water transportation. Notably, the distribution of these commodities helped create inland routes and the role of the itinerant peddler.

This interior view of a shoe shop is thought to have come from Portland. *Courtesy Maine State Museum.*

There were actually only a few pewter workers in Maine and they were active primarily between 1820 and 1850 in producing a new and improved pewter, Britannia, which was a shiny, less costly alternative to silver. Only twenty individuals were engaged in this manufacture in 1850, however, and the largest producer, Rufus Dunham of Westbrook, had a staff of ten and annually produced 4,000 teapots, 1,000 pitchers, 600 coffee pots and 1,000 gross of spoons, while nearby Goodrich and Porter produced 4,800 teapots.

Tin ware production was closely associated with pewter and led many to make both products. Goodrich and Porter, for example, produced three thousand tin pans in 1850 while also making Britannia teapots, and Belfast pewterer Samuel S. Hersey started his business making tin ware and he returned to this trade after only marginal success making pewter items.

The soldering of tin or the finishing of cast pewter was not very complicated and even one artisan working alone in a primitive workshop could easily outstrip the needs of customers in the local area. High per capita productivity was a characteristic of tin and pewter work and at mid-century a typical worker could produce some four thousand items per year, an average of thirteen items per day. Since tin ware was particularly inexpensive, the market tended to be strongest in the less affluent rural community which was most easily reached through the home delivery of the peddler's cart. Attendant to the centers of tin manufacture, therefore, was a network of highway cart traffic in which an exchange of communities often substituted for cash. Tin peddlers' wagons were a commercial expression of the realities of these manufactures.

This image depicts the Lewis A. Dutton sail loft in Boothbay Harbor, circa 1920. *Courtesy Maine State Museum.*

The first identified tin worker in Stevens Plains (now Westbrook) was Zachariah Stevens, who opened a workshop in 1789. Stevens not only formed a wide variety of containers, but he provided distinctive, decorative painting, which helped to identify the work of Stevens and his family and associates. In 1803 Stevens was joined by Thomas Briscoe—a man credited with establishing the first peddlers' route from Stevens Plains.

Straddling the major highway from Portland to Montreal, the peddlers of Stevens Plains were able to distribute tin ware and other commodities in New Hampshire, Vermont and Quebec Province and later, through other routes, to northern Maine, New Brunswick and Nova Scotia. As these routes became established, Steven Plains was a center of traffic, helping attract other tinsmiths to the area. In the first decades of the nineteenth century, Stevens and Briscoe were thus joined by the shops of Elisha and Elijah North and Oliver Dudley so that, by 1832, there were no fewer than eleven tin shops in Stevens Plains. Not surprisingly the community also produced brooms, combs and brushes that easily fit on a peddler's wagon and tin workers alone in that community produced upward of sixty thousand articles annually, even though the business had begun to decline under competition from silver plate. Artwork had a diminished value and what remained was the production of utilitarian products. As the era of decorated tin ware passed, the more usual products were pails, pots, pans and funnels, in addition to stovepipe.

In many communities, tinsmiths were the true manufacturers of stoves. In Hallowell, both A. Myrick and William Prescott made stoves, as did John Varnum of Readfield and Eben Jackson of Gardiner. In Portland Joseph Nash and O.S. Hartshorn offered stoves, while Erastus Kelly's Tinware Manufactory in Winthrop assembled three hundred cast-iron stoves and one hundred sheet-metal stoves as well as tons of stovepipe.

The rise of silver plating hastened the end of both decorative tin and pewter. Meanwhile, the more general distribution of hard currency slowed the bartering process and further diminished the role of the itinerant peddler. John Pennels, whose family had been involved in tin ware merchandizing at Stevens Plains since the 1830s, continued through the 1890s and was the last peddler to maintain a route. The sheds used in Stevens Plains to house carts and commodities were torn down around 1900.

Floor Coverings

Carpets

Carpets exported from Great Britain dominated the American market and this well-developed and centralized handcraft industry sent out Kiddeminster, Wilton, Brussels and Axminster carpets in quantities and at prices that discouraged local manufacture and even home production. Protective tariffs, imposed in 1824 and 1829, encouraged American manufactures while the increase in homegrown wool fostered local textile production generally, and carpet weaving particularly. Against this backdrop, carpet production came to Gorham in the late 1820s.

The arrival from Ireland of Hugh Gilroy, a "weaver by trade," launched this business in 1827. Using homespun raw materials, often provided from neighboring farm homes, Gilroy operated a small shop producing damask table coverings and "ingrain" carpets (also known as

Kiddeminster). Through Gilroy's efforts, Gorham became the carpet manufacturing center of Maine with a production of some 53,000 yards of ingrain carpeting in 1850, mostly produced by hand weavers and jacquard looms.

By the late 1830s there were already two shops at work in Gorham: the Gorham Carpet Factory and the Maine Carpet Manufactory. One of these, the Gorham Carpet Factory, was operated by Isaac Irish, who employed twenty-five to thirty operatives to make thirty-five thousand yards of carpet. His work was highlighted in the *Eastern Argus* on July 10, 1841, at a time when the business still exhibited many bespoke characteristics:

> *Farmers who wish to exchange their wool for carpets can do it on favorable terms; or by spinning their wool into yarn can have it manufactured to order, at prices varying from 37 to 55 cents per yard. Many farmers, I am told, and thus by increasing the quantity of stock put into their carpets, obtain a very firm and durable article...I am told that in order to have the yarn spun properly, directions should be received from the manufacturer. In a note from the Proprietor, he remarks to us as follows—"I do not hesitate to say that if our farmers will furnish me with their stock, which they can spin themselves, by having directions from me, that I can as heretofore give them entire satisfaction, by manufacturing for them as good an article as made in this country."*

During the years from 1830 to 1850, new inventions worked to revolutionize carpet weaving. The jacquard loom, first invented in eighteenth-century France, was employed in America beginning in the 1820s. While still hand operated, jacquard looms developed patterns by connecting each individual warp yarn to a control wire whose movement was controlled by punch cards similar to those later found on a player piano. Among the items used in the Gorham Carpet Factory in 1842 were "1 set punches for plates to cut cards," as well as "40 sets cards" representing 40 different patterns and some of these were 70, 72 and 75 card repeats.

The machine weaving of ingrain carpeting was perfected by Erastus Bigelow in Lowell in 1840, but this did not immediately transform carpet manufacture. At mid-century, Gorham's David McKinney was still employing twelve men and three women in hand weaving 24,000 yards of carpet and Daniel C. Emerby had a staff of twenty men and twenty women and products amounting to 28,000 yards. Power-driven machinery did hasten the centralization of the industry, however, and by 1860, Gorham had only one carpet producer. By 1870, this small-scale local production had ended.

Painted Oilcloth

Both imported and local wool carpets were relatively expensive. At the same time, the grip of merchants in the manufacture of floor cloths was never as complete as it was in the importation of wool carpets, and the role of skilled artisans was less essential. Together with the related trades of wall stenciling and furniture decoration, the production of floor coverings had a household tradition in Maine and saw the perfection of oilcloth and eventually modern linoleum. Though assisted by tariffs, the production of oilcloth floor coverings enjoyed a long, close relationship with local painting. Up to the middle of the nineteenth century, floor cloths were made primarily of

Sam Lancaster is shown here around 1880 working at his chair factory in New Sharon. *Courtesy Maine State Museum.*

cotton duck treated with numerous coats of paint. This provided a base for free-hand painting, stencil decoration or block designs.

Without doubt, a great deal of floor decoration in Maine homes was applied directly on the floorboards themselves. The commercial manufacture of oilcloths came in the 1830s and especially around 1840 when a number of floor-covering shops opened in central Maine. Within just a few years, oilcloth manufactories emerged in the towns of Hallowell, Manchester, Monmouth, Readfield, Skowhegan, Belgrade and Winthrop. In Hallowell there were two manufacturers at work by mid-century, and these employed thirty-six people to make nearly three hundred thousand yards of oilcloth. Meanwhile, in nearby Winthrop hand and horsepower were used to produce another three hundred thousand yards and eighteen thousand yards of oilcloth table coverings. In all, Maine manufacturers produced around one million yards of painted floor coverings by 1850.

Although oilcloth production was never highly mechanized, it still operated increasingly on a factory basis. Productivity demanded a wholesale distribution and, as usual, this led to centralization. Under the enterprise of the Bailey family of Winthrop, however, factory ownership took place in Maine and Winthrop soon boasted the largest oilcloth manufactory in all of New England and one of the premier manufactories in America.

The C.M. Bailey & Sons oilcloth factory began in the 1830s and absorbed most similar factories in central Maine, closing some while employing over two hundred people. This was good for Winthrop, but not always good for neighboring factory towns. In Readfield, one historian wrote bitterly that an industry that employed fifty people for a period of thirty years compelled them "to seek new employment or remove to new localities, and extinguishing the bright prospects of a thriving hamlet."

The Baileys purchased the Skowhegan factory in 1865 and in 1879 the activities at this site were reported:

> *There is situated in this village one of the branch houses of a very large oil cloth manufacturer. The well known C.M. Bailey, who is the king of oil cloth manufacturers in New England. Everything is done here by wholesale. Some 40 men are employed and the oils, paints and articles which go into the manufacture of these carpets are bought in very large quantities.*
>
> *The burlap which makes the foundation of the carpets is made of hemp, is purchased in Dundee, Scotland, and comes here by the ship loads in vessels owned by Mr. Bailey.*

The use of handheld stamping blocks helped maintain a sense of artisan manufacture even though factory settings replaced workshops at an early stage. The Baileys' operation was definitely a factory.

Bonnets, Hats and Shoes

Technology often transformed local workshops into centralized factories, making artisans into machine operatives. But in the production of bonnets, hats, tailor-made clothing and shoes, control of the raw materials often made them factory-like even while they retained the look and

John W. Penny of Mechanic Falls poses here with his patented Dirigo jigsaw around 1870. *Courtesy Maine State Museum.*

feel of workshops. In each case, mechanization was achieved primarily through use of the sewing machine and this tool was introduced simultaneously in homes, shops and factories.

Women's Hats and Bonnets

The milliner's trade centered on the manufacture and repair of bonnets and hats using silks, satins, ribbons, spangles and artificial flowers. In Maine, bonnets of straw were initially the most important of these and employed the most makers. Despite low material costs, high-quality straw bonnets could be expensive. These were made from plaits of six or seven ends of straw, braided by women in their homes, workshops and factories. This was reported in the Philadelphia publication *The Panorama* in 1837: "Great quantities of straw are plaited in families, especially in the New England states, sold to merchants, who, in turn, dispose of it to those who form it into hats."

Ambrose Colbey organized Colbey's Bonnet Rooms in Portland in 1847 while similar shops were established there by Hall, Johnson and Company before 1855. White's Bonnet Rooms produced a number of straw bonnets that were exhibited in Portland. The entry by E.G. Palmer was judged "one of the finest specimens of Leghorn straw we ever saw." By the 1880s the specialization of manufacture had created in Portland the trade of straw bleacher, but imports, primarily from Italy, ultimately eliminated bonnet rooms.

Men's Hats

As compared to straw hats and bonnets, men's felted and fur hats grew from a different and ancient tradition. The wool of sheep, llama and camel and the fur of beaver, otter, seal, muskrat and rabbits exhibited qualities that allowed the felting of individual fibers to form a coherent mass that could be formed into hats at the hatters' workshops. Here the hat bodies were formed around wooden forms, new fibers added and a surface nap raised through brushing. Although some machinery was introduced as early as 1847, the forming of men's felted hats continued as a workshop craft from the time of earliest settlement, and hats were still made to order as late as the 1870s.

In 1820 the total value of men's hats made in Maine was estimated to be $160,000, a sum nearly equal to the value of boots and shoes. Early in the decade, however, Maine hat making began to be centralized in urban areas, particularly Portland. In 1821 Bryon Greenough & Company was established and continued as a dominant enterprise until the 1880s, despite many shops continuing to offer works on order. A variety of furs and felts were offered, and in 1838 Greenough submitted to the Mechanic's Fair two "otter caps" using furs "taken in this state." In the 1830s other Portland makers included Martin Gore, Hiram Covell, John G. Hayes and Eben Webster. A division of labor was inevitable and so *The Panorama* noted that "the wollen bodies, used by hatters, are now [in 1837] often procured from persons who devote their attention exclusively to that part of the business."

The 1,500 dozen hats manufactured by Covell and Greenough in 1850 were, not surprisingly, produced from a supply of 1,500 dozen hat bodies. While small hat shops survived, specialization illustrated the direction of this industry. At mid-century, Maine's three largest producers, John

Workers pose near their lathes at the Fay & Scott machine shop in Dexter, circa 1920. *Courtesy Maine Historic Preservation Commission.*

C. Hayes, Byron Greenough and Covell and Greenough, produced over 20,000 hats and over 80,000 caps. A total of 103 people were employed, mostly women.

Shoes and Boots

The manufacture of bonnets and hats was restricted in both geographical distribution and employment, at a time when shoemaking was widely distributed in Maine and was of increasing importance to the economy. By 1850 the remarkable growth of shoemaking was obvious. While employment making hats and bonnets could still be counted in the several hundreds of workers, there were more than thirty-five hundred shoemakers enumerated in the census—a number of artisans eclipsed only by carpenters. Prior to the 1870s, this manufacture remained essentially a workshop trade carried on in homes, workshops and, occasionally, in small factories. There was no consistent chronology of growth and all stages of production existed simultaneously until the era of the Civil War when Auburn emerged as Maine's premier shoemaking center with as many as twenty-one factories, between two and four thousand employees and products amounting to over four million pairs of shoes.

Through the early decades of the nineteenth century, shoemaking followed a traditional pattern in Maine involving master, apprentices and journeymen workers. All tasks, from start to finish, except the tanning of hides, were commonly conducted in the cobbler's shop. Small, local cobbler shops provided custom shoemaking and finer shoes down through the 1860s. But the custom making of shoes was declining rapidly in the nineteenth century, giving way to regional,

This wool merchant occupied a storefront in Gardiner around 1880 and probably served both homes and factories. *Courtesy Maine Historic Preservation Commission.*

even national, commerce. If a Maine family wanted shoes in 1830, they might likely visit a general store where shoes were commonly available as pre-manufactured goods available for retail sale. The 1852 general store of Owen Doyle of Benedicta included in its inventory some 260 pairs of shoes while the Gardiner storefront of Michael Burke offered 850 finished pairs of shoes at mid-century. This was only possible because the custom work of the local shoemaker had given way to an age of mass-produced ready-to-wear shoes.

Specialization was, of course, an ongoing evolution. As with clothing, women's shoes remained specialty items long after the ready-to-wear fashions had been mass produced for men. The manufacture of women's shoes centered on eastern Massachusetts, particularly Lynn, and few Maine makers sought to compete for this market. Maine's shops and factories more often concentrated on boots, thick boots, brogans (ankle-high cowhide shoes with thick soles) and both light and thick shoes.

Mass production demanded the establishment of fixed sizes. After the hides were tanned, the first workshop step was cutting and sizing the shoe parts. This was followed by the assembly and sewing the uppers (the part of the shoe enclosing the feet and containing the tongue and lining), a process called closing and binding. This was the easiest step to master and, consequently, it was often farmed out to households and assembled by women and children on a piecework basis. These uppers were then tacked to a wooden last (wooden feet) along with the innersole and these were stitched (sometimes pegged) together. The outer sole and heel were then added.

There was work here for individuals in homes, in workshops and in factories. While the 3,500 shoemakers in 1850 were all men, an uncounted number of women and children were certainly involved. The quality of the finished products was, no doubt, variable. Up to the 1820, lasts were not often differentiated between left and right. Some very bad shoes were made and this fact led to considerable export to the American South, West, West Indies, Cuba and Haiti—markets dominated by the three million slaves in these areas. A wholesale trade in boots and shoes grew rapidly in the Winthrop area in the 1830s and 1840s. Joshua Wing, Isaac Nelson, Joshua Trufant and S. Johnson Philbrook were among the shoemakers who had workshops at their houses along Monmouth Road and employed several men each. At first, most of the shoes were sold in New England, then Boston. Isaac Nelson sold his goods directly to New Orleans while, after 1849, the Wing brothers shipped a large part of their products, and those of their neighbors, to California.

Shoemaking was a growth industry in Maine during the 1840s and 1850s, and it was not uncommon to find as many as twelve or fourteen shoemakers at work in small shops in one town. In Norway, for instance, the partnership of J. and E.W. Howe commenced in 1853 and immediately employed nearly thirty men and women in a rented shop. It was noted in 1853 that

> *Shoe making seems to be the order of the day in the village. J. and E.W. Howe are doing a pretty smart business, employing 100 hands, male and female. Mixer and Watson are quite smartly engag'd. M.P. Hall are preparing to erect large shoe shops next spring…shoe making is flourishing, about 70 cases of Shoes and boots are made per week.*

The mass production of men's shoes, and the mechanization of manufacture, accelerated beginning in the 1870s and this change brought more shoe shops to Maine.

III.

The Mill and the Furnace

There are 5,151 streams in Maine represented upon the State map. These ramifying into innumerable branches thread the surface of the State with a fine network of brooks and rivulets, so that in all parts it seems alive and in motion with running water.
Water Power of Maine, 1869

Power sources were largely incidental to artisan workshops, but in mills and furnaces power was central as they cut logs into boards, ground corn into meal, carded woolen fibers for spinning, finished woolen cloth to make it serviceable or ground rags for papermaking. Mills and furnaces were ultimately single-purpose processing plants that took one raw material (for instance, logs) and turned it into another raw material (boards). Mills and furnaces did not interfere with artisan trades; they did not compete with domestic workers and rarely produced consumer goods. For example, the final product of the gristmill was cornmeal, not bread. Sawmills never supplanted cabinetmakers even though they revolutionized the work of the hand sawyers. The role of the mill and furnace was to assist the manufacturer of consumer products, not to supplant them.

The integration of power and process was so complete in many mills that it was difficult to separate the mill structure from the water system that drove it. Up to the eighteenth century, saws and grindstones were commonly connected directly to water wheels. The concept of moveable machines was a nineteenth-century novelty that evolved first in textile factories. Since waterpower sites were location specific, mill industries were located accordingly. Mill privileges filled up with industries that served both local and commercial interests.

The abundance of waterpower, its broad distribution and its proximity to timber resources and navigation conspired to make mills the foremost of all industrial concerns in nineteenth-century Maine, and there were over 3,000 "mill privileges" counted in 1869. These generated some 2,656,200 horsepower; the equivalent of "34,000,000 men, laboring without intermission."

The Technology of Waterpower

From 1630 to 1830 there was little change in waterpower technology. Only four types of wooden water wheels were used: undershot, overshot, breast and tub. Using both gravity and

This unidentified photograph shows two hand sawyers cutting boards. Powered sawmills were placed everywhere in America from the earliest days of settlement. Labor was scarce and timber relatively plentiful. Whipsaws, like that shown here, survived into the nineteenth century in shipyards, where they were used to cut irregular shapes. *Courtesy Maine Historic Preservation Commission.*

momentum, the related physics were unchanged. Since water seeks sea level, a body of water held higher than that was a form of potential energy that might be used, in its descent, to turn water wheels. Calculations of both height and water quantity determined the amount of work that could be done and, at the same time, determined the best kind of water wheel for the job.

The builders of Maine's first mills knew that they could regulate power with dams and flumes as well as gates and water wheels. But these tasks required an expenditure of energy and a cost that did not always seem worthwhile when the resource seemed limitless. Even very small, seasonal power sites used inexpensive flutter-type undershot wheels despite the waste and the ephemeral nature of many such wheels. In Wales, Maine, for instance, one mill was described as being "on a brook, where some little sawing is done, during the spring snow meltings, and during the season of freshets." Noticing this predisposition, one chronicler, writing in 1749, noted that "New England abounds in saw mills of cheap and slight work...These mills mostly stand upon small streams, because cheap fitted...[but] they do not afford water sufficient to drive the wheel in summer." There was great tolerance of waste—both of timber and of waterpower and virtually all types of water wheels were employed. Little initiative could be seen to improve engineering and the largest waterpower sites went unused anyway because the work was too costly and difficult.

Interest in waterpower engineering came to Maine in the 1830s and 1840s as wooden breast wheels were replaced by iron and steel turbines in the region's textile mills. Wooden wheels were rapidly considered obsolete and the manufacture of portable iron wheels moved the work inexorably from the millwright to the foundry. The styles of these wheels included percussion, centre-vent, reactor, scroll and spiral vent. The United States Patent Office issued some three hundred patents for turbine designs and improvements in the late nineteenth century. By 1870 there were many turbines used in Maine, with the most successful being the Reynolds, Rose and Tuttle wheels. Four (the Carleton of Garland, the Gould of Skowhegan, the Blake and later Holmes of Gardiner) were made in the state.

Sawmills

The technology of sawmills was not developed in Maine or even in North America. Parts for Maine's first two sawmills (destined for South Berwick and York) arrived in New England in July of 1634 aboard the vessel *Pied Cow.* The "forest primeval" stretching north and inland from here made sawmills a commercial cornerstone in the New World that it could never have been in the Old. The first explorers had hoped to find gold; instead, they found trees, and lots of them.

The romantic story of the Maine woods has dominated the state's history and it is easy to imagine that Maine's lumbering led the nation. It did not. But Maine's dominant position on the eastern frontier assured that those activities were magnified in both reality and legend.

The attraction of commerce, particularly timber and lumber sales, was an incentive that promoted settlements: sawmills paralleled closely this new population. During the eighteenth century the harvesting of timber resources worked its way northward through York, Kennebunk, Saco and Gardiner. In a common pattern, the sawmill built in Bethel in 1774 by Joseph Twitchell preceded the first frame house by five years. Even the virgin forest had its owners, of course, and the holders of patents, or grants, were sometime possessed of timberland they never actually visited. From the earliest time there was a mosaic here of petty fiefdoms. Logging and lumbering and the sale of merchantable boards were ubiquitous Maine industries that were widespread in their community influences.

Maine's earliest grantees wasted little time devising ways to extract the "gold" of their wilderness acres. In the seventeenth century, York County was the principal center of sawmilling in the Province of Maine and this region continued to serve as the major outlet for sawn lumber in the eighteenth century. Maine's first documented sawmill was established at Asbenbedick Falls in the present town of South Berwick (then part of Kittery), and later known as "Great Works." And at the same time another mill was erected on "old mill creek" in the town of York. In addition to these pioneers, the search for timber led the industry steadily northward and eastward along Maine rivers; before 1660, sawmills were built in Wells, Saco, Casco and Falmouth and, by the end of the century, there were more than sixty mills along the coast from Hampton, New Hampshire, to Casco, Maine. The greatest number of Maine mills

This *Sketch of Mechios Mills*, printed in 1776, is the earliest and among the best images of Maine's eighteenth-century up-down sawmills. *Courtesy Maine State Museum.*

was in Kittery and up the Piscataqua River to the present towns of Eliot and South Berwick as well as nearby York and Wells.

Emphasis shifted to the Saco River during the half-century from 1770 to 1820. There were no fewer than seventeen mills located "above the falls" at Saco, and more stretching upriver to Hollis, Buxton, Bar Mills and Limington. Many mills here were built before the era of large log drives and these ventures were often characterized by individual initiative and small partnerships. Accordingly, many sawmills were owned in fractional parts as small as one twenty-fourth or even one forty-eighth. Both the mill and the "mill privileges" were sometimes divided in small shares. Thus, one owner of the "Proprietor's Mill" in 1806 sold "one tenth part and one forty-eight part of the mill 'privilege,' and also one forty-fifth part and three thirtieths of a certain mill privilege [on the] western side of Saco River in said Limington…together with one forty-fifth and three thirtieth parts of a double saw mill standing on said privilege."

Down the river, at Bar Mills, several saws were perched on a bar in the Saco River, and these too were divided in fractional ownership. Furthermore, the lower mill had two saws, the stream saw and the shore saw, and these were also owned separately. When Samuel Sands purchased a portion of this mill in 1807, his deed included "one fourth part of the stream saw in a double saw mill called the lower bar saw mills." It was also noted that "the shore saw of said mill is now owned by John Palmer and Joseph Atkinson." This free-for-all was somewhat regularized by

Photographer Aaron Houdlette of Dresden caught this image of the aging sawmill, erected by Sylvester Gardiner in 1754. This was one of the first mills built along the central Kennebec. *Courtesy Maine Historic Preservation Commission.*

translating fractional shares into days of work, and thus when Daniel Cleaves sold his share of the upper stream saw at Bar Mills in 1817, he conveyed "five days or five twenty-forths on the steam saw in the upper Barr Mill."

A very different tradition unfolded on the Kennebec River in eighteenth-century Gardiner. Here, beginning in the 1760s, Dr. Sylvester Gardiner founded an empire based on control of waterpower and sawmilling. There were absolutely no fractional shares of any mill built here; rather, the policy instituted by Sylvester Gardiner—sustained until the 1830s by his grandson-in-law and heir, Robert Hallowell Gardiner—was to hold title to all the dams, mills, buildings and water rights and lease them to tenants. Here was a mini-empire that controlled countless mill sites and over one hundred thousand acres of land through merchant control unparalleled in Maine prior to statehood. Along the Cobboseecontee Stream in Gardiner there were no fewer than thirteen mills on "dam #1" alone. Generations of sawmills were built and rebuilt here. In the future, Maine lumbering was destined to be composed partly of individuals working alone or in small groups in inland locations and, at the same time, large mills established north of Gardiner and particularly on the Penobscot.

The story of logging and lumbering on the Penobscot has long dominated the history of Maine's lumber industry and created the "romance" of the Maine woods. Although the Penobscot was sparsely populated with sawmills prior to 1830, this drainage had become the

As the period of small water-powered mills ended, the picturesque quality of these scenes appealed to photographers. This photo of the old stone mill was taken by Aaron Houdelette of Dresden, circa 1890. *Courtesy Maine Historic Preservation Commission.*

dominant cornerstone of sawmilling in Maine by 1850. The sawmills above Bangor in Orono, Veazie and Stillwater produced most of the 176,000,000 feet of the state's long lumber produced at mid-century.

The commercial trade in lumber dominated the statistics and the story of Maine sawmilling—sawmills do not produce cities, young single men do not give rise to permanent settlements and, consequently, neither do lumberjacks and sawyers. Today the lumber towns of Veazie, Orono, Oldtown and Stillwater are still small towns. Commerce produced the city of Bangor. The exploitation of the Penobscot region grew rapidly beginning in the 1830s and by 1842, one hundred million feet of long lumber was sawn each year. The *Bangor Republican* was happy to note that "Maine furnishes about three-fourths of all the white pine lumber exported from the United States and the Penobscot River is the center of this trade." Earlier, in 1836, the *Niles Register* had claimed that

> *There are now in operation night and day, within a few miles of Bangor, principally within the limits of Orono, more than 200 sawmills for boards…manufacturing considerably more than 1,500,000 feet of boards daily. This is cutting lumber on a scale which no one, who has ever been "down east" have an adequate conception of.*

Henry David Thoreau's view of the "Maine Woods" in his book of the same title recounted that in the villages of Oldtown and Stillwater the mills were built "directly across the river." Of Bangor he wrote most poetically:

There stands the city of Bangor, fifteen miles up the Penobscot, at the head of navigation for vessels of the larger class, the principal lumber depot on this continent with a population of twelve thousand, like a star on the edge of night, still hewing at the forest of which it is built…and sending its vessels to Spain, to England, and to the West Indies for its groceries.

From the seventeenth to the mid-nineteenth century, the technology of sawing changed very little. Expansion was shown by the number of sawmills in Maine. By 1790, there were 150 sawmills in the province, but thirty years later, in 1820, there were some 750 and by 1849, the census counted over 1,400 and some of these were equipped with multiple saws. There was also a vast archipelago of mills that functioned and served a smaller, even local, region.

When Penobscot County had eclipsed York and Kennebec as the state's lumber emporium at mid-century, there were counted 3,111 sawyers plus a very large number of laborers and loggers.

The journal of Ezra Beal provides a glimpse into the organization of smaller sawmill enterprises in Norway at mid-century. In 1854 Beal owned the mill but hired Augustus Wilkins of Waterford to

run the saw mill the present season and until all the lumber now at the mill and the pond is cut into, taking one half of the proceeds of the saw mill, shingle and lath mills to himself. He is to find his tools, files, etc. and cut lumber, relieving the owners of the mills, or myself, from care or responsibility concerning the saw mill, charging for such services as is done not belonging to running the mill and machines and properly taking care of the staff. Mr. Wilkins agrees to keep the mill and machines in repair as far as can be done with his own skill. The smith work to be paid for by Deal. He is to run the mill at all the time when the water can be spared from the grist mill and cut the lumber to such dimensions as the owners of the logs may direct. He is to cut up and throw out all slabs and firewood not suitable for laths and take—of the wood to himself.

The Technology of Maine Sawmills

The first sawmills were really water-powered versions of hand-powered pit saws. By hand, boards were cut by a two man crew: the sawyer above who toed the mark, lifted and guided the saw down the length of the log, and the pitman below who pulled the saw down and made the cut. In England, where timber resources were scarce and labor plentiful, this practice was commonplace. In America, sawmills flourished in a world turned upside down with much timber and few workers. From the 1630s through the 1840s, the principal technology was the up and down saw, a powered version of the hand-operated pit saw. The water wheel turned a crank that lifted and then pulled down on a pitman to move the straight blade, around seven feet long. This blade was held in a wooden framework and tipped back slightly (the kilter of the saw) so that a cut was revealed with each upstroke. In this space the log was inched forward to prepare for the next down stroke. As compared with modern-day sawing, this work was painfully slow, resulting in only about one to two thousand feet of lumber sawn each day. But, as compared to hand pitsaws, these mills were marvelous improvements that operated at speeds suited to undershot water wheels. There was no rush to change.

Advancing technology often led the way to new techniques of manufacture suitable to machinery but that differed from any non-powered precedent. This was true of sawing boards and, not surprisingly, the idea of cutting lumber using a continuously acting circular saw occurred in the eighteenth century. A quantum leap in speed was necessary and this was often provided by faster moving iron water turbines. The circular saw dominated the development of mills during the second half of the nineteenth century and this was in concert with the faster moving water turbine. Turbines and circular saws advanced together and sawmills became cheaper to erect and portable.

But this technology actually sustained dispersal and encouraged the production of portable sawmills throughout Maine. There was a potential for centralization and, indeed, this can be seen in the Bangor area particularly. In this region, millers most often preferred to use greater speed and power provided by ganging together a larger number of straight saws. Thus, the history of sawmilling in Maine does not exhibit a singular sequence of growth and change. This reality helped sustain both the largest and smallest mills, both a profusion of portable circular mills and gang saws stretching entirely across the Penobscot.

Manufactured Building Materials

As compared to the sawing of long lumber, it is easy to overlook the production of products like clapboards, shingles and staves. It is logical, however, that manufactures of sawmill machinery to cut long lumber also offered saws to produce these smaller items. The spread of mills for these was directly related to the development of specialty saws for each of these. Until the second quarter of the nineteenth century, many of these products were split and shaved by hand through the work of laborers or farmers. Sawing machinery often meant the transfer of these manufactures from the farm outbuilding to a water-powered site. Productivity was greatly increased while labor decreased. It was estimated, for example, that a hand shingle maker, a shingle weaver, could produce one thousand shingles on a good day. Using saws, the *Hallowell Gazette* reported in 1856 that two men in Durham had sawed and bundled eleven thousand shingles in one day. But eight years earlier, a record thirteen thousand shingles were sawn in Swanville to win a ten-dollar bet.

The sawing of laths from the sawmill "waste" of slabs was truly ancillary to the sawmill itself, but the other products were largely independent manufactures, though often associated with the waterpower of sawmills. In 1870, for instance, there were just 300 shingle mills in Maine, while another 128 of these saws were listed as part of the state's sawmills. Clapboards, for example, were sawn radially from bolts cut to the length of finished clapboards (usually four feet) and shingles were sawn from bolts around sixteen inches long. In either case, preparation of stock for these saws differed from the sawing of logs into boards, and they were usually produced apart from sawmills and not always ancillary to them.

The production of shingles was a piecework process (shingles were sold by the bundle, or count) and speed mattered. The use of shingle saws spread quickly and there was considerable competition to make improvements. Augusta machinist A.G. Johnson soon offered "Johnson's Improved Shingle Machine" in 1847 and this innovation received extended notice in the

Scientific American of that year. By mid-century, Maine was sawing nearly one hundred million pine and cedar shingles each year and, not surprisingly, the centers of this work were in Orono and Gardiner, where much of the total output was produced. Meanwhile, many mills offered clapboards as a sideline and by the end of the Civil War there were as many as fifty clapboard mills operating and over two million clapboards reportedly made in Orono alone.

Yet another group of woodworking machines was introduced in the first half of the nineteenth century: planers and shapers. Planing machines allowed for the preparation of lumber produced in standard dimensions. This revolutionized building construction and architecture itself as the balloon frame replaced the earlier post and beam techniques. Through action of the planing machine, the two-by-four was born. This machine encroached on the craft of carpentry and made the production of standard forms a function of sawmills, not carpenters. The production of doors, sash, blinds and later ornamental woodwork followed quickly. Many mills not serving the commercial distribution of boards became emporiums for the production and sale of home building products. Some of the larger door and sash mills were in Kennebec County, and in the decade of the 1860s the firm of Jeremiah Furbush of Waterville employed a suite of eleven machines (including three for planing and two for mouldings) to produce 10,000 doors, 60,000 sash and 2,600 blinds. In Augusta, D.W. Mosher used fifteen machines to make 12,000 doors and a large quantity of sash and blinds while J.P. Wyman made 18,000 doors, 14,000 sash and 7,200 window frames. House building now started with millwork of standard dimensions and common, reproducible parts that were no longer custom made. The introduction of planing and shaping machines, as well as drill presses and band saws, altered mill industries in a fundamental way. Meanwhile, though, specialty saws for clapboards and shingles did not alter the mill industries from which they grew. Under the influence of portable woodworking tools, some mills began to overlap the work of the artisan workshop and evolve toward a true factory.

The expansion of lumber mill enterprises raised them (and many smaller mills) to a position only exceeded by the production of long lumber, while the processing of commodities at the mill became common.

Gristmills

Sawmills most often preceded the spread of agriculture and population but, necessarily, the construction of flour and gristmills followed. Farming was often a consequence of spreading population, but rarely an incentive. Nevertheless, the vast majority of Mainers called themselves yeomen and, later, farmers. Regardless of profession or craft, the ownership and tillage of land had a particular appeal that suggested home ownership and independence. Farming, not mechanical work, fit the image of the America envisioned by many founding fathers. ("Founding Mothers," remembering their worn-out mothers, might have held a different view.)

Gristmills emerged as a direct service to the needs of the local population and they were followed by other mills that served the local agricultural community. These included fulling and carding mills in addition to grist and flourmills.

With little commercial incentive, there were already 524 gristmills in Maine in 1820 and this was second only to the number of sawmills. While the sawing of lumber exceeded the needs

This 1860s view shows the waterpower of Norway along a stream where Ezra Beal had leased a mill site in 1850. *Courtesy Maine Historic Preservation Commission.*

of the population, the grinding of grains was less than its needs. Rather than provide local services for farmers, merchant grain mills often supplied products in support of textile-making communities and provided ship provisions. Accordingly, gristmills often displayed a pattern of distribution that was similar to that of the blacksmiths: small and local. There was a limit to the distance a bushel of corn could be carried. Unlike blacksmith shops, though, gristmills were somewhat expensive to build and, as a consequence, they were often held in fractional parts similar to Saco River sawmills. The work of grinding corn was not thought to require much skill. For this reason this work was consigned to young men who were commonly employed to tend the mill, not to own it.

Even where the flow of water was small, harnessing the total fall of water could be substantial. A series of dams was sometimes needed and this was the case near a Norway mill shown in this photo taken in the late 1860s. *Courtesy Maine Historic Preservation Commission.*

Throughout the entire period of the early seventeenth through the late nineteenth centuries, the grains found in Maine grist and flourmills remained the same and so did the grinding technology itself. The work of local gristmills was heavily weighted toward the grinding of cornmeal. Corn, which had been discovered in the New World, was the most plentiful grain raised in Maine by a wide margin, and Indian corn, so called, amounted to 1,750,000 bushels in 1850, against only 300,000 bushels of wheat.

Corn growing was decentralized and so were the gristmills, which commonly did their work "on toll," with a share of the meal given over to the mill as payment. Meanwhile, wheat was nearly as important and while this production lagged behind the state's actual needs (imports were common from the south and west), the sale of tools for this process (grain cradles, flails and winnowing baskets), followed by fanning mills and, later, threshing machines, all reflected a substantial use of wheat despite the heyday of corn. Additionally, the farmers of the state grew substantial quantities of rye, buckwheat and oats and all of these except oats were commonly ground into meal, flour or fodder in local grist and flourmills.

The effort to separate the grain kernels from the husks or chaff was intermediate between the harvesting of grains and the grinding processes. These tasks were often done by hand or with farm machinery when it was available. In the case of corn, the kernels were removed from the cob after it had been dried. This separation was sometimes accomplished by scraping the cob with knives. During the early decades of the nineteenth century a variety of corn shelling machines was available. Racliffe's patent corn sheller, first developed in the 1830s, was manufactured by several Maine makers.

Wheat required a process of threshing to separate the germ from the chaff. In the eighteenth and early nineteenth centuries, this was simply done by beating the grain with sticks and then tossing the mixture of the two into a slight breeze to blow away the lighter husk. This serendipity was replaced with the hand-cranked fanning mills and, later, by threshing machines.

The technology of grinding grains remained nearly the same through the middle of the nineteenth century. In the typical gristmill, the grain was fed by gravity through a hole in the upper of two stones, or runner, that was turned by the water wheel below. This stone turned against the lower, fixed stone, the nether, while a furrow carved into each stone worked the ground grain from the center to the outside. These two stones were covered by a wooden box (the hoop) and the finished product emerged at the bottom. The greatest technological change here was in the speed and efficiency of the water wheels and turbines themselves.

Grinding corn was faster and easier than grinding wheat, and while some mills were capable of grinding three hundred or more bushels of corn each day, the grinding of wheat rarely exceeded four bushels per hour.

Flourmills

Although the milling of wheat was undertaken "on toll," commerce led to merchant involvement at an early date. Large flouring mills controlling this commodity were developed in the middle states as early as the eighteenth century, and Maine imported a great deal of flour from this region, even though the state embarked on its own flour milling industry. In Gardiner, William M. Vaughan invested over $25,000 in a brick mill that had been built by Robert Hallowell Gardiner in 1844 and he produced over fifty thousand bushels of wheat per year.

The operation of merchant flourmills was more complex that the operation of a custom gristmill. While the latter offered the service of grinding, the former was engaged in the sale of a manufactured finished product. The most important ancillary machinery was for bolting, which sorted the flour into different grades. Vaughan's mill, for example, produced in 1850 some 12,000 barrels of flour, 65,000 barrels of midlings and 28,000 bushels of bush feed. This grading was important but well beyond the capacities of most local gristmills.

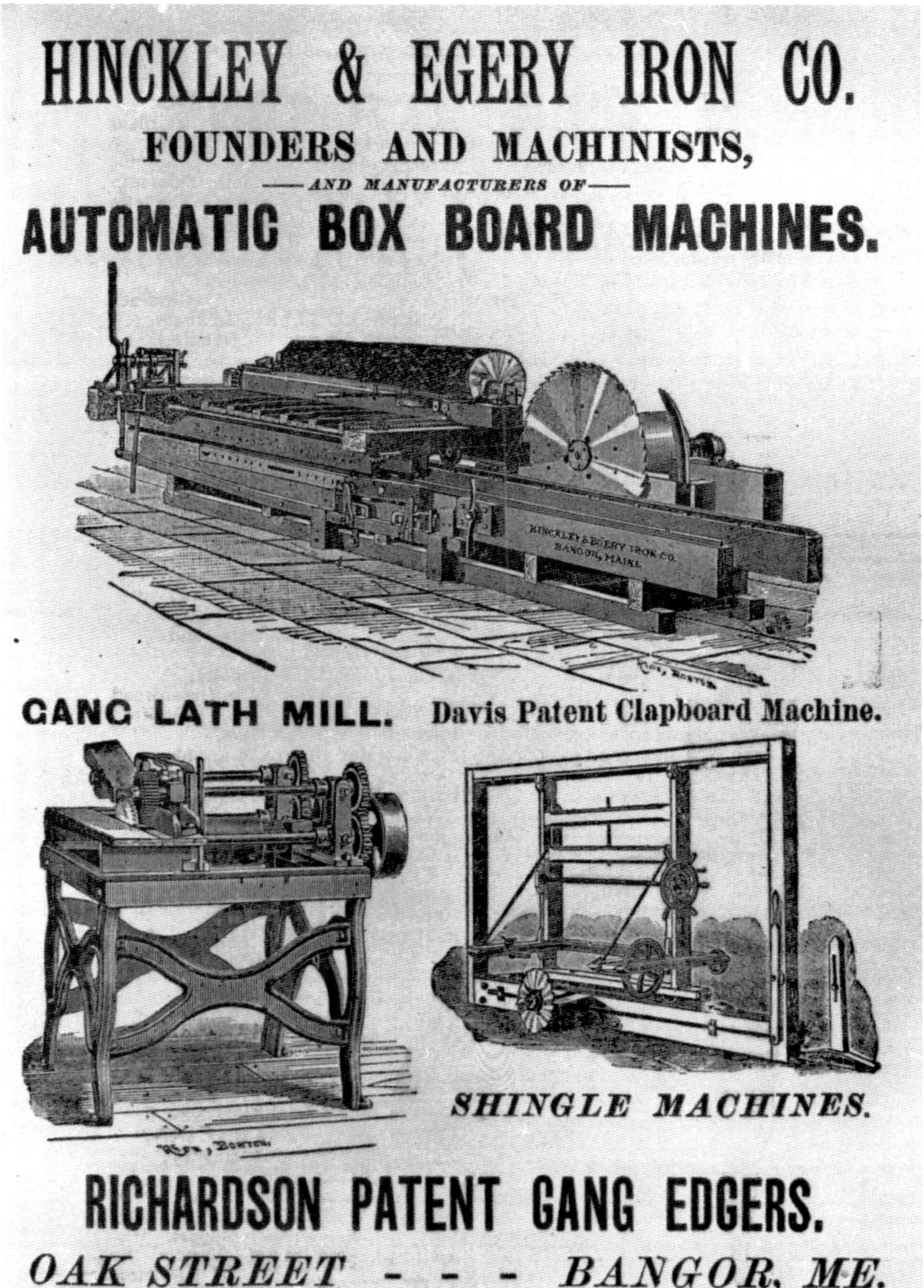

Hinckley & Egery of Bangor advertised a variety of ready-made saws. Shown here are a clapboard mill as well as a larger circular saw for boards, barrels and lath, as illustrated in the *Maine Business Directory* in 1885. *Courtesy Maine State Museum.*

Portable mills led to temporary conditions, as evidenced by this Sheepscot-area photograph taken by E.J. Leighton around 1910. *Courtesy Maine State Museum.*

Increasing flour production in Maine centered on the inland communities of central Maine—particularly Somerset, Franklin and Oxford Counties, where 50 percent of Maine wheat production was raised in 1850. Larger flourmills were built in Lewiston and Auburn, as well as Kingfield, Freeman and Strong. The largest of these was operated by William King, who at mid-century had ground over four hundred thousand pounds of flour. Capital needs led to a centralization of this enterprise while, at the same time, rural gristmills remained quaint countryside survivors. Gristmills resisted efforts at centralization and slowly they became obsolete even as flourmills surged forward.

Canning

Nathan Winslow of Westbrook is credited with perfecting a system for hermetically sealing tin cans as early as 1829, a byproduct, no doubt, of Westbrook's tinplate production. The widespread use of canning was not seen in Maine, however, until the 1870s when it signaled important changes in the American diet. Corn canning became an important industry in Maine in the 1880s and this, in turn, fostered the planting of sweet corn. Large canning companies were established in the 1880s, and by 1890 numerous packers were producing at least 150,000 cans

An anonymous work crew posed for this Fred Libby photograph at this portable mill in the Sanford area around 1910. *Courtesy Maine State Museum.*

of corn per year. During the busy season, the Portland Canning Company employed some 150 men, women and boys to fill some 300,000 cans of corn.

Canning signaled some important changes in the American diet that altered the course of farming and the processing of grains. Since that time, corn has been directed to canning plants or used mostly as cattle feed. The "age of cornbread" in New England and Maine had passed.

Fulling and Carding Mills

In terms of dollar value, it was claimed early in the nineteenth century that the production of textiles in the home was the largest business in Maine. Seventy-five years later, textile manufacturing was the state's largest factory industry, employing the greatest number of people. As elsewhere, this signaled an enormous change in manufacturing strategy that, from the late eighteenth century onward, has been called the Industrial Revolution. The change from home to factory was incremental, however, and often slow; this was an evolution more than a revolution. This evolution saw simultaneous work in the home, the artisan workshop, the mill and the factory. The role of mills was intermediary between the home and the factory; mill power was used to

Large-scale factory systems impacted the architecture of sawmills, as seen here in a view of the double sawmills of Arthur Berry and N.O. Mitchell of Gardiner taken by A.F. Morse of Hallowell around 1870. *Courtesy Maine Historic Preservation Commission.*

lighten and quicken home manufacture, serving the needs of hand spinning and weaving at the same time that it led this work away from the home to the factory.

Mill, furnace and kiln industries were often consolidated into factory organizations in the nineteenth century. Therefore, sawmills tended to be part of the manufacture of doors, sash and blinds, while gristmills tended to be associated with commercial bake houses and casting furnaces associated with the production of stoves and architectural iron fences. Tanneries enjoyed a long association with the manufacture of boots and shoes, but until integrated in a factory setting these remained single-purpose mills and shops using the waterpower and the natural materials as found.

When Maine became a state in 1820, there were already 149 fulling mills and 200 carding mills supporting tasks in the home. These two mills, often associated with each other, were, in fact, distinct and when working together they constituted the third most common use of waterpower. Fulling mills had existed in New England from the first decades of settlement; the technology was medieval in origin and quality woolen cloth simply could not be produced in the home without the services provided by the fulling mill and clothier. By contrast, the carding mill housed a machine process that did not appear in New England until the 1790s.

Fulling Mills

Woolen cloth often exhibited a coarse surface texture when it came directly from the loom. At this point, this cloth was, technically, a flannel and lacked the serviceability and appearance of high-class woolen fabric needed for clothing and blankets. Through the process of fulling, damp fabric was worked under heavy hammers to produce felting by partly unweaving the cloth while shrinking and thickening it. This required the use of waterpower and thus this process had been little changed since medieval days.

Weaving was a home process, but high-quality cloth also required the work of a mill and an experienced clothier—a combination of waterpower and artisan skill. Here, mill power was coupled directly with the work of the clothier, whose efforts raised the surface with brushes and teasels, cut the surface nap with heavy shears and dyed and pressed the cloth to produce a uniform, plush finish. These were artisan trades in short supply on the New England frontier, and this tended to distinguish the clothier's work from that of sawing and grinding. The potential of the clothier to ruin a piece of cloth was great and the quality of workers was, no doubt, variable. Some, including David True of North Yarmouth, were clothiers on a part-time basis. More typical, though, was the work of Daniel Carleton of Alna or Joseph Yetton of Peru, who practiced their craft in either a new facility or the old fulling mill. Since waterpower and mill buildings were costly, they were often built and rented by merchants, not craftsmen. James Conner of Unity, for instance, owned the local fulling mill and waterpower, but relied on Joseph H. Hill to carry on the business from 1814 to 1819 and, beginning in 1820, he leased the mill to two clothiers, Messrs. Soule and Cram, who agreed to "take and keep the said mill until the expiration of one year."

Faced with the paucity of qualified clothiers, America took the lead in the development of shearing machines to replace the skilled hand shearing. Many inventions, including that of Liberty Stanley of Winthrop, led ultimately to the perfection of helical shear, similar to the reel-type blade of a lawn mower. The first shearing machine used by Daniel Carleton in Alna was known as a vibrating shearing machine, a dead-end technology made by Parks and Davidson in Vermont shortly before 1830. Despite the limitations of this machine, there was an inexorable search to replace traditional craft skills with the machinery.

Fulling mills provided their services at the end of the home weaving process and they rarely made a successful transition from custom service to factory organization. Domestic hand weaving of these products entered a steep decline as power looms for woolen broadcloth were perfected in the 1820s and after. As late as 1850 there were still in Maine over two hundred men who called themselves dressers/clothiers, but there were, in fact, only twenty-six active fulling mills, and all of these were in conjunction with carding mills and machines Typical was the mill/shop of Dimon Furnel of Wilton who operated his shop only sixty days a year and pressed slightly over two thousand yards of woolen cloth.

Carding Mills

Carding mills, on the other hand, provided their work at the beginning of the home spinning and weaving process. They too began by providing services to the home but carding machines occasionally led directly to woolen manufacture. The carding mill was most often a service to home spinners. These services diminished or eliminated factory work.

Individual wool fibers exhibit a rough, barbed surface and yarns can be made with the use of friction among these barbs, in addition to twisting. Woolen fibers are teased, rather than combed, and carding was essentially a process of brushing to produce a rolag of evenly distributed fibers ready for spinning. This was a common home process that used hand cards until the early decades of the nineteenth century. These hand cards employed a series of closely set wires and through a slow, laborious process these cards were worked against each other to tease wool fibers into a long strip ready for the hand spinner.

Carding was a bottleneck in the home production of yarn and efforts to mechanize this process began in eighteenth-century England. The technology of machine carding was copied from the British in the 1780s and brought to New England by 1790, along with mechanics from England and Scotland who could build and operate them. Maine remained on the cutting edge of carding improvements when the technology was brought to Maine in the early 1790s by Samuel Mayall and the story of Maine's first carding mills is inexorably tied to this family. When Mayall arrived in Gray around 1791, there were no carding machines in the province; but in the years ahead, carding mills were established in Farmington as early as 1800, and in Limington, Saco, Lisbon and other locations from 1804 to 1809. In these locations, carding machines provided custom services to home spinners as advertised in the *Eastern Argus* by John Mayall in 1805:

> *Wool Carding Machine*
> *John Mayall, informs the inhabitants of Saco, and the neighboring towns, that he is now erecting a machine for Carding Wool at Saco Falls; which will be ready to Card by the 20th of June. Price for Carding, Eight Cents per pound; mixed Wool Ten Cents, cash. The Wool must be picked and greased, or those who prefer to bring it ungreased, may have it Oiled for Two Cents per pound. Those who favor him with their custom, may Depend upon having their work well executed, as he has served seven years at the business, and is not one of those who pretend to follow a business they have not learned.*
> *John Mayall*

By 1820 there were 210 carding mills in Maine and the period from 1800 to 1820 might be called the "golden age" of home manufacture. Under the influence of rapidly growing woolen mills, the production of woolen cloth moved dramatically from the home to the factory after 1820, and by mid-century there were only 20 carding mills left in the state of Maine. Carding machines were now the opening sequence in the production of the woolen factory itself and no longer independent supporters of home manufactures.

Other Water-powered Mills

Water-powered mills were used to cut, stamp, pound and grind a variety of useful products ranging from bone meal used as fertilizer, oakum used in shipbuilding, excelsior for packing and upholstery and even gunpowder. Of these "lesser" products, the most important were probably the grinding of hemlock bark for tanning and the shredding of textiles for paper.

Paper

Maine's prominence in papermaking came toward the end of the nineteenth century under the influence of new papermaking processes using ground wood pulp as a principal raw material.

This photograph, circa 1880, is looking toward the western side of Bonney Eagle Falls and shows a double sawmill that was equipped with two single saws in 1869. At this time, this sawmill was capable of producing some two million feet of long lumber each year. *Courtesy Maine Historic Preservation Commission.*

Prior to this 1860s innovation, paper was commonly made using cotton and linen rags and Maine's role was not particularly striking. In the earliest paper mills, rags were beaten to a pulp using waterpower to drive ironclad hammers sometimes termed a stamping mill. Using an eighteenth-century machine, the hollender, iron blades helped cut rags into smaller pieces.

The spread of papermaking in Maine, and indeed in all of America, was handicapped by a lack of skilled paper makers and a shortage of rags. Both of these handicaps plagued a 1730s Falmouth mill, the first in Maine. It was reported in 1733 that Samuel Waldo and Thomas Westbrook's pioneer paper mill "had not begun to work for want of materials," and this was to prove an ongoing problem.

Between 1800 and 1850, however, the number of paper manufacturers in Maine increased from virtually none to over seventy. The products of these mills tended to be writing paper, but more often newsprint and other cheaper commodities, and one mill in North Yarmouth produced wallpaper. The quantity of work was not impressive, except that here alone, among mill industries, women played a leading role. Norton and Davis of Hamden, for example, employed twelve men and twelve women to manufacture 132 tons of newsprint using 200 tons of rags. Richards and Hoskin of Gardiner used twenty-one women and thirty-four men at mid-century

Here Rufus Dwinell replaced earlier double sawmills with a large block of saws that spanned the river. Each of these saws was leased. *Courtesy Maine Historic Preservation Commission.*

in the handling of 600 tons of "rags and sundries." The shortage of rags led to experimentation with other component materials, including straw and wood and even, in the case of one Gardiner importer, Stanwood and Tower, the purported use of "Egyptian mummies."

Bark and Tanneries

The contribution of the forest to wood product manufacture is obvious; more easily forgotten is the contribution of hemlock trees to the manufacture of leather. Tanning is a lengthy, laborious process in which the principal raw material, in addition to hides, is the tanning agent—hemlock bark in the Northern states.

Tanning was mostly a hand process except for the grinding of bark. Because of this grinding process, tanning is usually considered a mill industry. Grinding was a simple process of cutting, pressing and pulverizing that was sometimes possible using horse power. As late as 1850, horses still produced bark mill power by walking in a circle in the towns of Falmouth, Westbrook and Gardiner. While bark mills could exist separately from tanneries, this was an exception. In the nineteenth century, tanneries were the bark mill's greatest customers.

Leather was properly ranked as a necessity and, indeed, a vast array of commodities depended on this material. Accordingly, tanneries, along with saw and gristmills, often led the settlement of Maine. By 1810 there were already 200 tanneries in the district and this number grew to 248 in 1820 and to 395 by 1840 toward a high point of production in the 1870s.

The tanning process was long and unpleasant; it was a job that was not for the squeamish or impatient. Careful attention was important and technological change was limited. There were four essential operations: first was a preliminary washing process for the skins, lasting about thirty hours. This was followed by a process of soaking and scraping, which loosened the hair and fat and required soaking in lime vats for one year. These skins were then placed on a large horse, or beam, where the tanner used fleshing knives and scrapers to complete the cleaning process. The hides were then soaked in a bark solution for twelve to eighteen months. In all, two to three years were required for the entire process.

Hemlock grows throughout Maine, but until the middle of the century this was not an incentive for the location of tanneries. Rather, there was a close relationship between leather and the users of leather, who were the true manufacturers. The relationship of tanning to shoemaking was especially notable, and this likely was a decisive factor in the geographical distribution of tanneries. The first tannery in the town of Leeds, for example, was established by Samuel More in 1814 and continued until after mid-century through the production of tanned hides and shoes. Meanwhile in Paris, Moses Hodgdon was known variously as a farmer, tanner and cordwainer. As late as 1850 Benjamin Smith of Houlton employed three men in the tanning of seven hundred hides and another twelve men in making two thousand pairs of shoes and boots. A Harrington tannery produced two thousand hides in 1850 while fifteen men labored in the production of some eight thousand pairs of boots.

Timber cutting of hemlock trees was secondary to bark production and, in an age dominated by pine, large stands of hemlock were simply wasted. Much of Maine's hemlock supply had been used up by mind-century and, accordingly, the search for bark became a reality that determined the location of tanneries. During the period from 1840 to 1870 there was a shift toward the north, toward the state's hemlock lands. At the same time, there was a process of enlargement and centralization. The Shaw brothers nearly cornered the marketing of leather in the last decades of the century, with the *Somerset Reporter* noting in April of 1879 that "in this land of hemlocks we find an abundance of tanneries, especially in the eastern part of the State. The Shaw Brothers, who own several large tanneries in this State and Canada, virtually controlled the leather trade in this country."

Sea lanes and inland waterways assisted Maine's commerce in tanned hides. In 1878, for instance, the tannery in Skowhegan imported some fifty thousand sheepskins from South America. With the exhaustion of hemlock stands, Maine lost its advantage and, meanwhile, centralization led to the decline of tanneries in the state by the end of the century.

Kiln Industries

The raw materials essential to all kiln industries are relatively cheap products of the earth—limestone, clay and sand—and the cost of fuel and transportation was often greater. The characteristics of these simple materials can be seen by reference to the production of lime, bricks and pottery. These raw materials, together with abundant firewood, yielded mortar, plaster, bricks, pipes, tiles, pottery, glass and bricks.

Lime

The most substantial of Maine's nineteenth-century kiln industries was the burning of limestone rock to produce lime. Limestone is itself composed mostly of dead sea organisms forming a concretion of calcium and magnesia (calcium carbonate). This is then burned in order to produce lime (calcium oxide). Lime has the special property of slaking with the addition of water and it was thus a critical ingredient in making plaster, mortar and cement. Centered in mid-coastal Maine, particularly Thomaston (and that part of Thomaston that became Rockland), Maine's lime industry grew to national proportions. Lime from this part of Maine virtually dominated the New York City market for mortar and plaster.

The burning of limestone was started on the coast by William McIntyre in 1733 and, until nearly the present, this region has been dominated by the story of lime and its ancillary needs for wooden barrels and firewood. In the eighteenth century these kilns were often little more than heaps of stones placed on other heaps of firewood (flare kilns). By 1800, structures began to take on definite forms created by walls of fieldstone and granite. Pot kilns were in use by the 1830s and these featured clay-lined interiors and mortared exteriors over fieldstone and granite, and by the first decade of the nineteenth century there were thirty-five kilns in the area around Thomaston. But the process itself was little changed and while draw kilns used less firewood, it was usual that each charge required thirty cords of wood to produce forty-five tons of lime after round-the-clock tending.

Bricks

Brickyards generally accompanied the expansion of Maine's population, and particularly the growth of commercial and industrial buildings. The widespread presence of clay deposits was an incentive to manufacture bricks, but not a motive. Houses in Maine were usually built of wood and the use of bricks tended to be limited to the chimney, hearth and foundation. The need was basic, but not compelling, and the largest users of brick were not the builders of homes. Accordingly, the need for bricks grew rapidly in the decades of the nineteenth century under the influences of urbanization and industrialization as the builders of large factories turned to brick as a preferred building material.

High-quality clay deposits existed throughout the state, but production was promoted by nearby projects. In 1850 there were 389 brick makers listed in the census and most of these were clustered near factory cities and had arrived from 1830 to 1850, which included the building of both Saco and Biddeford and the cotton mills and machines shops rising along the Saco River. Not surprisingly, there were seven brickyards in Saco and five in Biddeford, together burning between 6 and 10 million bricks per year. When the development of Lewiston followed from 1860 to 1870, brick production increased from around 800,000 to over 3 million each year. The building environment spurred coastal trade in bricks and Maine became a substantial exporter. Arch Morrill, Gardiner brick maker, has been credited with sending large quantities of brick to Boston while in Edgecomb some eight brickyards employed 30 men and burned 1.5 million bricks each year.

The undisputed center of brick export was Brewer where, at mid-century, there were no fewer than twenty brickyards together producing and shipping fifteen million bricks each year that,

This rare view from Head Tide shows the interior of one of the state's many small gristmills. This postcard view probably dates from the 1870s. *Courtesy Maine Historic Preservation Commission.*

along with lumber, were sold at auction in Boston. As with most other kiln industries, the cost of raw materials was secondary to the cost of handling and transportation. Brewer's yards used three thousand cords of firewood and employed men to tend the fires, day and night, for nine days per burn. Here, as elsewhere, there was no standing kiln; rather, the bricks were assembled in a kiln form, plastered over, burned and the entire kiln torn down to retrieve the finished brick. Then the labor-intensive process of rebuilding the kiln began again.

There were many more brick makers in Maine than there were potters. But bricks remained anonymous products while the work of potters was distinguished by style, color, workmanship and materials. Since potters produced recognizable consumer goods, their products can be identified and collected today.

Pottery: Redware and Stoneware

Two types of pottery were familiar in Maine homes: earthenware (or redware) and stoneware. Both were made in Maine. The earliest of these was "redware" and these were produced from local clay deposits that were essentially free for the taking. Its name came from the color of the fired items, which resembled present-day clay flowerpots. Because these wares were relatively soft and porous they were usually glazed with a harder material.

This view showing the interior of the Hapgood carding mill was produced at Center Lovell in 1902. *Courtesy Maine State Museum.*

This rare vibrating shearing machine was made around 1830 and used at the Carleton Woolen Mill in Alna. The bolt boards on the floor nearby are also from this mill. *Courtesy Maine State Museum.*

Redware potteries existed in the state as early as the 1730s and initially were sometimes produced as a joint effort with bricks and lime, as was the case of the Norcross pottery in Hallowell in the 1790s. However, most potteries were distinct from other kiln manufactures. These included those of James Bickford, who began his pottery in Buxton in the 1790s, and Benjamin Dodge in Fayette around 1800, Daniel Holmes in Alfred in 1804, Ebenezer Corliss in Yarmouth in 1806, Samuel Philbrick in Skowhegan in 1808 and John Alld in Hollis in 1811. These first redware makers produced a fairly large range of domestic containers including pots for cream, brown bread and beans; jars for molasses; jugs for cider and vinegar; and milk pans. Eventually tin, glass and stoneware took shares of this market and, by 1850, many redware potteries were producing just a few staple products. John H. Stafford of Monmouth, for instance, produced 225 dozen milk pans, 100 dozen cream pots, 100 dozen bread pans, 75 dozen bean pots as well as "other ware" in 1860. Meanwhile, another Monmouth potter, Silas H. Coburn, made 2,650 dozen milk pans, 80 dozen bread pots and 40 dozen bean pots.

Without doubt, Monmouth could not absorb and use 5,700 milk pans each year and assuredly these products were sold in large lots to retail merchants. But under pressure from tin, glass and stoneware, the popularity of local redware declined. The *Yankee Farmer* in Cornish noted in 1835 that

> *We have found most all kinds of ware generally used among us for milk, cream, butter, and many other purposes, are liable to many objections, and we have wished to find some better article that would be cheap, durable, and pure. We find that stoneware has all these advantages.*

Unlike redware, stoneware could not be made from Maine soils. Clay for this stronger, harder product was imported from New Jersey, and glazing clay from upstate New York. While redware required around thirty-six hours of firing, stoneware required some six days. Coastal trade was demanded to provide the supply of raw materials and this tended to centralize the production in urban, commercial environments. While making redware might have been promoted by clay deposits, the production of stoneware grew directly from the needs of local farm families.

Maine's first stoneware manufacturers did not appear until the 1830s and the importation of stoneware continued from the beginning of the nineteenth century until the launching of the Bangor Stoneware Company in the 1880s. Until that time, production of stoneware was confined to Portland and Gardiner. For nearly a half-century, a succession of manufacturers (lasting on average six years) managed a production that was more or less continuous. Portland, Martin Crafts (1834) and Eleazer Orcutt (1835) combined in business until 1837, when they gave way to Caleb Crafts & Company. A few years later Clough, Calhoun & Company was replaced by Winslow, Clough and finally J.T. Winslow. In Gardiner, Alanson P. Lyman established Lyman & Clark in 1837 but this pioneer firm was replaced in 1839 by R. Thompson & Company, and this firm was replaced by Ballard and Brothers. F.A. Plaisted followed in 1855 and continued until formation of the Gardiner Stoneware Company. The location of these potters adjacent to rivers and streams was not an accident when imported raw materials were essential.

Eventually, of course, the advance of mechanization diminished the role of the local potter who had hitherto thrown each vessel one at a time on a hand wheel. Beginning in 1850 the J.T. Winslow Company introduced machinery to replace handwork and, as it did, molding machinery replaced circular forms with pitchers and architectural shapes. Mechanization was rapid after the Civil War and hand-thrown pottery was obsolete and even downright antiquarian, as noted by the *Bangor Commercial* in 1916:

> *A demonstration that is attracting large crowds is that of the newly-opened store of the Bangor Stoneware Co. at 74 Central Street. One of the potters of the company is engaged in making various kinds of crocks, pots, and jugs, using for that purpose an old potter's wheel. This is a distinct novelty to many, and the interest of the crowds gathering outside the window is unflagging.*

Furnaces for Iron Ore

In 1800 woodworkers were the true manufacturers of items like plows, while tinsmiths assembled stoves. Iron and steel became the dominant materials in the industrial economy in the nineteenth century as forges, furnaces and machine shops produced most everything from textile machines to steam engines. Despite predictions that Maine was destined to erect major iron furnaces, investors saw that Maine was not positioned well to benefit from this.

Bog iron ores were indeed mined in Maine as early as the eighteenth century and shipped to furnaces in Boston and Plymouth. There were five ironworks built in Maine before 1800, including the Forge Steam Ironworks established by Jeremiah Powell in North Yarmouth around 1755 and Jeremiah Moulton's ironworks, built in Sanford in 1772. The construction of blast furnaces in Maine continued to be slow in the nineteenth century. Perhaps typical of iron furnaces in Maine in the eighteenth and first third of the nineteenth centuries was the Shapleigh Iron Company, which was started in 1837 in Newfield by a number of investors from Portsmouth, New Hampshire. When "in blast," these furnaces were charged with a mixture of bog iron ore, charcoal and limestone (or seashells) for flux. This furnace was charged about twenty times a day with four boxes of iron ore and ten bushels of charcoal, along with eight bushels of clamshells. The iron produced was drawn from an arch at the bottom and allowed to run in sand troughs into oblong shapes called pigs. No consumer products were produced and this raw material was marketed to others.

The mining of iron ores, and the construction of blast furnaces, was limited in Maine and might easily be overlooked altogether if it were not for the Katahdin Ironworks. Erected near a small mountain of iron ore (known variously as Iron Mountain and Ore Mountain), the Katahdin Ironworks followed the Pleasant River Iron and Steel Company and the Maine Iron Company. The ore here was abundant and inexpensive and waterpower was available to drive the furnace's bellows and firewood was at hand to make charcoal. Geologist Charles T. Jackson estimated that it would last for thirty-one years and since the old-fashioned furnace used charcoal, Jackson thought:

This wide-ranging view shows the furnace and many ancillary buildings of the Katahdin Ironworks. *Courtesy Maine Historic Preservation Commission.*

> *The Iron made at your furnace is of excellent quality especially for machinery...being soft grey iron easily turned or planed. It's superiority to Iron made with coal or coke will give it preference in the market over English and Pennsylvania Irons which are made with coal.*

But the site had disadvantages and these included the fact that it was a full fifty miles from Bangor and the shipping lanes of the Penobscot. Thus, the Katahdin Ironworks violated two cardinal needs for iron furnaces: favorable geographical location and adequate labor supply.

The process relying on charcoal was, indeed, superior but also archaic. The production of charcoal used up 1,200 cords of wood each year and required between three and five hundred woodcutters seasonally.

The furnaces, roasting ovens and charcoal kilns of the Katahdin Ironworks were unique in Maine history; the Iron Mountain site has stood like a solitary outpost in the Maine forest. No other place in Maine was quite like it.

Rolling and Slitting Mills

Pig iron was of little use to village blacksmiths who needed a more malleable product to fashion into horseshoes, tire irons, plow points and nails. It would be romantic to think the pig iron produced by furnaces like the Katahdin Ironworks was turned into consumer items in the hands of village blacksmiths, but this was not the case. Blacksmith items were made from wrought-iron bar stock in which heat and hammering had further oxidized the iron, a refining process reducing carbon content. Most wrought iron was, in fact, imported from England, Germany and particularly Sweden and Russia. The importation of bar iron was so extensive that it dominated the marketplace and discouraged production in America, including Maine. The refining process could be managed directly at the blacksmith's hearth, but water-powered hammers were more often employed. This work often resembled a mill more than a furnace and was often, in fact, distinct from the ore furnace itself and could thus be located nearer to strategic routes of commerce. Bar stock was often worked into useable shapes and sizes in rolling and slitting mills, adjunct to the ore furnace, which used waterpower. These mills were often distinct, producing flat iron plates with rollers or nail rods cut to standard sizes, new raw materials provided to the local blacksmith or, more frequently, of reproductive manufactures, of cut and rolled iron products. Like ore furnaces, rolling and slitting mills were relatively scarce in Maine.

The Saco Nail Factory

In the seventeenth and most of the eighteenth centuries, pounding heads on nails was done by hand using cut nail rods. Nail cutting machinery was invented in New England, and naileries used furnaces to reheat sheet iron plates and used waterpower to drive cutters and rollers and portable nail-making machinery. The Saco Nail Factory was started in 1811 by Thomas Cutts and Josiah Calef, who were the first to employ the valuable waterpower site on Cutts Island near the mouth of the Saco River. Here, commerce delivered raw materials (Swedish or Russian iron) and distributed the products of a nail factory described as "among the most complete in the country." Employing twenty-five men, the Saco Nail Factory made a large number of nail rods as well as 100 of the largest nails per minute and 150 shingle nails per minute, toward a total of some 3,500 pounds per day by 1830. These nails were mostly sold in Boston and used in America.

The Pembroke Iron Company

The first failed effort to establish an iron refinery and rolling mill began in Pembroke in 1828 and *McLanes' Report on Manufactures* reported in 1832 that

> *a very extensive establishment, for the manufacture of bar iron, is being erected at the new town of Pembroke, in the county of Washington, which will be in operation in the course of the season, employing 500 men, 3 ships to bring scrap iron, etc., and about a dozen other vessels to transport wood, etc.*

This grand scheme failed, as did the effort started in 1844, but finally, in 1849, the Pembroke Iron Company was formed and it enjoyed some thirty years of prosperous business. The company was distinguished from the Saco effort by its plan to produce wrought iron from pig iron. The experienced labor to do this was not available in Maine and, in fact, the state's contribution was limited to waterpower, sea routes and firewood; otherwise there was no compelling reason to locate this facility in Pembroke. Pig iron generally came from the Lake Champlain region; ore came from Lake Champlain; labor came from England and Ireland; and the products were shipped to Boston and to Taunton or Douglas, Massachusetts, or New Haven. By 1850 the company was operating a fleet of vessels while the furnaces consumed 3,000 tons of pig iron, 1,000 tons of iron ore (to aid in the oxidizing process) and 600 tons of coal. The result was some 2,200 tons of bar iron and nails, hinges, shafting, anchor chain and other products.

The Katahdin Ironworks, the Saco Nail Factory and the Pembroke Iron Factory were unique in the history of Maine. This was not the case, however, for casting furnaces.

Casting Furnaces

Until sometime around 1700, cast-iron products were produced in Maine by taking molten iron from a blast furnace and forming specific shapes in sand molds. The production of pig iron allowed development of less expensive cupola casting furnaces, not necessarily located near ore deposits, which re-melted scrap or pig iron for local use. While there had rarely been more than one blast furnace at work in the state, there were many dozens of foundries, a total of twenty-five at mid-century, turning scrap and pig iron into consumer products and machine parts.

The casting process was suited to the mass production of identical parts and this provided an advantage over the blacksmith shop. In Maine, this capacity is illustrated by the manufacture of stoves, agricultural implements and machinery. In each of these cases, the foundry furnace first played an ancillary role as an independent supplier of cast parts used by tinsmiths, plow makers and machine builders who were the true manufacturers of these products. By 1850, though, foundries were increasingly located in conjunction with woodworking and machine shop activities, and the products were frequently finished machines and consumer products. The foundry often emerged as the true manufacturer. This is illustrated by the production of stoves. At the beginning of the nineteenth century, these consumer products were made by tinsmiths using stove plates from foundries that were frequently in Pennsylvania. During the second quarter of the century, however, the focus of stove production shifted from the workshop to the foundry itself, and local furnaces often replaced imported plates. As early as the 1830s, Wood, Bishop & Company of Bangor produced thousands of stoves. By 1850 the Estes and Whitten Foundry produced some three thousand "cook and other" stoves. Wood & Bishop marketed its Clarion line of stoves until the end of the century, and the Portland Stove Company produced its Atlantic ranges into the twentieth century.

When machines were made mostly of wood, foundry work was supportive, but ancillary. Mass production demanded identical interchangeable parts and the precision made possible by iron and steel technologies. Machine building in the nineteenth century came to be dominated by the machine shop and the foundry. Together these advanced and promoted the use of iron and steel. In Maine, this work included the manufacture of textile machines, machine tools and sewing machines, as well as the technological leaders of the age, locomotives.

Mill, furnace and kiln industries were often consolidated into factory organizations in the nineteenth century. Therefore, sawmills tended to the part of the manufacture of doors, sash and blinds, while gristmills tended to be associated with commercial bake houses and casting furnaces with the production of stoves and architectural iron fences. Tanneries enjoyed a long association with the manufacture of boots and shoes, but until integrated in a factory setting, these remained single-purpose producers standing alone using waterpower and the natural materials as found.

IV.

The Factory

Operatives of the best character and in any required number, can be procured from amongst the sons and daughters of our farming population; accustomed to labor, robust, trustworthy, and intelligent and glad for employment within their native State at moderate compensation.

Water Power of Maine, 1869

The term "factory" does not have well-defined boundaries. In the eighteenth century the term "manufactory" was used to designate an enterprise making goods by hand. In the nineteenth century, ironically, it came to mean the opposite: an enterprise dominated by machinery. Furthermore, the search for productivity led many industries to evolve into factory organizations. Thus the manufacture of wooden grain cradles ultimately became the Dunn Edged Tool Company, a factory producing thousands of scythes and axes. To make matters even more complicated, many factories retained the names assigned to them during their first iteration. Cotton and woolen mills, shoe and machine shops, were most often factories and virtually every product of Maine manufacture was, at some point, a true factory.

Many relatively small manufactures were organized on a factory basis, and these used separation-of-labor concepts and very often machinery. For the purposes of this research work, a factory is defined as a manufactory providing a large, centralized employment coupled with product specialization suited to a national marketplace. These characteristics best distinguished true factories from the shops and mills alongside or preceding them. In Maine there were many that fit this definition, but four exerted a great, altering impact on the state's economic history: cotton and woolen mills and shoe and machine shops. These four hoarded the vast majority of the state's capital investments and they also produced many population centers seen today. Some of these enterprises, notably textile factories, were drawn to Maine by the state's waterpower opportunities. Others, like shoe manufacture, were promoted by relatively inexpensive labor. Shoemaking came relatively late in the nineteenth century, but the factory manufacture of woolen and cotton goods led the way toward the American factory system. In Maine, too, textile manufacture produced social consequences changing demography and economic structure. It is the study of these enterprises that best defines the meaning of the factory in nineteenth-century Maine.

Cotton Manufacture

Both linen and wool were employed in the Maine home in greater quantities and much earlier than cotton. But it was the manufacture of cotton cloth that introduced factory organizations to the state. Although many raw materials were not produced in the state, cotton mills were the largest factory employers and they produced many centers of population found in the state to this day. Despite a few earlier mills, large-scale development was seen first in the Saco/Biddeford area and, after the loss by fire of an earlier wooden factory under construction, cotton production was launched in the summer of 1832 using substantial capital raised by Massachusetts investors. By that time, however, Samuel Slater's first cotton spinning mill in Pawtucket, Rhode Island, was more than forty years old and the important prototype factories had been built in Waltham, Massachusetts, twenty years earlier and perfected in the city of Lowell, Massachusetts, more than a decade earlier. The York Manufacturing Company of Saco followed a now-familiar development pattern, but the cotton industry actually enjoyed a domestic, and later a workshop, phase from the eighteenth century forward.

The production of cloth became regimented, standardized and mechanical before it was actually mechanized. Even though cotton yarns were produced in America beginning in 1790, cloth was produced in family homes for at least twenty-five years more. Even if woven in many different places, standardization of designs, weights and appearance was needed if merchants were to market these products broadly as other than custom, local work. Yarns were put out to be woven in Maine homes. Yarn weights were standardized by measuring the number of skeins needed to make one pound, which equaled the yarn count. Goods were uniform and constructed to industry standards. All cotton factories produced essentially the same catalogue of goods and they were necessarily often in competition with each other. As compared to the manufacture of woolens, the roles of the garment makers and designers were not so important.

Sample books were assembled showing the use of specified yarn colors to produce stripes and checks following weaving instructions from the yarn mill. Rhode Island yarn mills marketed yarns and cloth throughout New England, including Maine, and even though power looms had been introduced at Waltham beginning in 1814, there was little immediate impact in Maine. Despite use of power looms in Brunswick in 1820, it was estimated that some two-thirds of cotton cloth was still being woven on handlooms at the time of Maine statehood.

The 1820s was a decade of transition. Plain goods could be easily produced with power looms while a variety of checks and stripes continued to be woven by hand. Maine merchants sold a variety of yarns, usually from No. 6 to No. 22 (skeins per pound) and, at the same time, sought to retain home weavers to make cloth for sale. The impact of Waltham's power looms was not immediately felt. In 1822 Elias Kelsey, who had long offered a variety of homemade cloths to Portland area consumers, published a "Notice to Weavers" seeking to hire "immediately 150 GOOD WEAVERS to whom constant employ and fair prices will be given." The following year Nathaniel Crockett advertised for "100 or more GOOD WEAVERS" and John B. Cross wished "to employ immediately 200 good WEAVERS." In 1823 Henry Brody offered large quantities of cotton yarn, cotton cloth of many descriptions and Waltham woven cloth, while he also advertised for "3 hundred Good WEAVERS."

The mills built in Saco during the 1830s were funded by the Boston Associates and were copies of the successful cotton mill built by members of the associates in Waltam, Lowell and along many of New England's major rivers. *Courtesy Earle G. Shettleworth Jr.*

Merchants like these carved out a niche in the local economy as both retailers and middlemen manufacturers. The centralization of the cotton factory was about to end all of this; the 1820s proved to be the last gasp of the put-out system.

The great choices among cotton products came through printing. Power looms introduced at Waltham and used by the Maine Cotton and Woolen Manufacturing Company in Brunswick in 1820 produced plain goods such as the 100,000 yards of sheeting made by one hundred employees of the Brunswick factory. Printing designs on this plain cloth—calico—was an important spinoff industry ultimately leading to new factories developed at Lowell and Taunton, Massachusetts, and Dover, New Hampshire. The production of cotton cloth was started in Maine in 1827, when quantities of cotton cloth woven by power looms in Dover were taken overland to Springvale, where blocks and tables were used to make a variety of printed designs. These calico goods were sent down the Mousam River from Print Works Falls to Kennebunk. The printing of lightweight cotton prints was an extremely important spinoff industry that dramatically increased the range of cotton products.

The construction of the York Manufacturing Company in Saco was a watershed event for Maine. During the late 1820s small spinning mills were virtually eliminated in favor of a more systematic development following examples previously demonstrated in Lowell, Massachusetts, and Somersworth, New Hampshire, as well as other locations. Investors had learned that the high capital cost of developing manufacturing sites was best managed by the development of a number of separate corporations to build dams and canals, to erect manufacturing buildings and to produce the machinery needed for the several factories.

These two unidentified weavers posed for a photograph around 1875. *Courtesy Maine State Museum.*

Investors sought appropriate sites with ample power and access to transportation. The river systems of Massachusetts and New Hampshire had filled up first with clones of the Lowell system. The Saco River near tidewater was easily the most noticeable of Maine's waterpower opportunities. With a river system beginning at the White Mountains, the Saco River had an ample and reliable supply of water, which had been used mostly for the sawing and shipment of lumber (some twenty-one million board feet produced as late as 1827), but this was a finite resource that was beginning to run low. The most favorable of all sites was on Indian Island, later called Cutts Island. Here Thomas Cutts and Josiah Calef of Boston had built a rolling and slitting mill for iron rods and nails as early as 1811 and this had emerged as the state's largest iron working industry.

By 1850 no other Maine community came close to this region in matching the concentration of investment, buildings, waterpower and people. Cotton manufacturing was, of course, also underway in a dozen other communities, including North Berwick, Oxford and Augusta. Though important to the economic expansion of Maine, these activities were all small in comparison to the works in Saco. The scale of investment, the cost of machinery and the competition of the marketplace worked to lead cotton factories increasingly to merchant leadership.

In 1826, the Saco Manufacturing Company was organized expressly to build a cotton factory on the site. The organizers of this development apparently believed that the Lowell factories had succeeded because they were large. Consequently, investors set out to build a factory of truly massive proportions. The building, a seven-story wooden structure measuring 210 by 47 feet, was completed in 1826 and was clearly the largest cotton factory built in America up to that time, nearly twice the size of factories being built in Massachusetts. Unfortunately, wooden construction was not the answer and predictably, the entire factory burned to the ground in 1830 just as cotton manufacture was getting underway. This ended the brief era of large wooden cotton factories and also ended the fortunes of the Saco Manufacturing Company. This cash-poor company ran aground yet again in the manufacturing depression of 1829. It was replaced in the early 1830s by the York Manufacturing Company.

At first the interests of the York Manufacturing Company embraced all of the related aspects of land ownership, waterpower management, construction of buildings and the production of machinery. It has already been found useful, however, to separate the task of cotton production from the building of dams, canals, buildings and machinery. The Saco Water Power Company soon followed to develop factories and mill sites as described in an 1844 circular:

> *The Directors…have determined…to proceed forthwith to build a Cotton Mill, to be carried by the Water Power Company, unless a new Company, exclusively for manufacturing purposes, should take up a portion of the power employ the Water Power Company to erect its buildings and machinery. This latter course is for them desirable, as keeping the appropriate business of the Water Power Company distinct from manufacturing.*

In anticipation of the Water Power Company's offering, the Laconia Company was incorporated, and it contracted to buy the needed mill site on the opposite branch of the Saco River in Biddeford and to secure an option there for more sites if needed. In addition to all the machinery, the Laconia Company required the building of a mill and three blocks of tenements

The Pepperell Mill in Biddeford employed primarily a French Canadians labor force. Company threats did not keep many workers from returning to Canada to help their families at harvest time. Ultimately, the mills simply closed during this period. *Courtesy Maine Historic Preservation Commission.*

as well as an assortment of outbuildings, and this was repeated again a month later for the construction of a second mill. In addition to the four mills of the York Manufacturing Company, the Water Power Company built mills for the Laconia Company and others soon assumed by the Pepperell Manufacturing Company.

Between 1831 and 1850 factory buildings, tenements, offices and boardinghouses continued to be built with little respite. The Saco/Biddeford area soon became Maine's first manufacturing city, transformed from a town of lumbermen and fishermen to a city of brick factories. The York factories in Saco alone were weaving over six million yards of cloth per year, using over 750 looms and employing some 250 men and 1,000 women! The opening of several Pepperell Manufacturing Company factories a short five years later quickly allowed Biddeford to boast of

Many families living in the Pepperell Manufacturing Company's ten tenement buildings referred to them as the Ten Commandments. *Courtesy Maine Historic Preservation Commission.*

its eleven factories, operating a total of 100,000 spinning spindles and employing 3,000 men and women. There was made here some thirty million yards of cotton cloth each year.

Developments in Saco/Biddeford were watershed events for Maine. They also document a period of dramatic technological change for the whole American textile industry. Samuel Slater had worked to create the first American cotton-spinning mill in Pawtucket, Rhode Island, in 1789 and 1790—a good forty years before new efforts to develop began for a second time in Saco. In the late eighteenth century, machinery in both Great Britain and America was made with cast- or forged-iron parts attached to a wooden framework These machines were relatively small, portable and each designed to perform one task in the sequence of producing yarn. From that start, machinery became increasingly more sophisticated and it moved faster. This evolution was inexorable but gradual. The pattern of smaller, moveable machines was sustained by machine

A spinning room is pictured here in Lewiston, circa 1880. *Courtesy Maine Historic Preservation Commission.*

shops. By the time that the Saco works were being planned, cotton factories remained the point of organization for these machines—where they were laid out and operated.

The motivating forces for all of this were water wheels, which in the 1830s were usually still made of wood and located in the shelter of the basement. These wheels, normally breast-type wheels running using the weight of the water, provided power to the factory upstairs through a series of line shafts running the length of the factory. This whole mechanism moved slowly, probably around forty revolutions per minute, and it was cumbersome and heavy. Each machine was driven through these shafts and a series of gears. Factories were commonly planned to produce a specific product with the complement of the various machines needed to permit continuous production. The work of managing these machines was often limited to the movement of the finished products to the next machine in the sequence. Ideally, there were no bottlenecks, no slow-ups or delays. In practice, however, the productivity of some machines outdistanced the capacity

Shown here is the weave shed of the Continental Mill in Lewiston, circa 1875. *Courtesy Maine Historic Preservation Commission.*

of the next. Thus, the work of moving stock and materials from machine to machine was fitful rather than precise. It was a classic case of "hurry up and wait." For labor, this meant the need for long hours with many stops punctuated by frenetic periods of activity.

There were many changes underway by the middle of the century. American manufacturers, led by Rhode Island's Zachariah Allen, had shown that speed could be substituted for weight and size. Thus lighter, faster-moving line shafting could substitute for heavier, slower-moving predecessors. Machines could now run faster, driven by pulleys and belts rather than gears. At the same time water turbines were engineered to replace wooden breast wheels and these generated more power and more speed. Meanwhile, the width of factories, usually forty to forty-five feet maximum, was governed by building designs and pent roofs. The length of the line shafting was limited by the torque of the line shafting. But more power meant that more of these factories could be built on the same power sites, and more machinery driven faster. Machine shops inherited the pressure of faster spindle and loom speeds, and the somewhat relaxed pace of early cotton millwork was replaced by speed. A visit to the Saco factories at mid-century still would have shown a sequence of smaller, portable machines. By now, though, iron and steel were the favored building materials and faster line shafting the general rule. The quality of life was declining as speeds and productivity increased.

From the 1830s through the early twentieth century the number of employees working in Maine cotton mills continued to grow, but after mid-century the number of factory sites was not enlarged—with the notable exception of Lewiston. Here, the period from 1850 to 1870 saw the city rise from a rural country town to the foremost manufacturing city in the state. As had been true earlier in Saco/Biddeford, the capital needed was raised primarily in Boston.

Lewiston

By the time of Lewiston's development, the American cotton industry had reached a mature phase, both economically and technologically. Lewiston was not an experiment. This city's turn came relatively late, but the factories built here were among the largest and most modern in New England.

The waterpower of the Androscoggin River was among the best in Maine and, as late as the 1840s, it had scarcely been used for any manufacturing purposes, though the potential had been obvious for decades. In the 1830s a number of leading citizens, notably Edward Little, sought to develop the Androscoggin's power and in 1836 he established the Great Androscoggin Falls, Dam, Locks and Canals Company with the dream of cloning the success of Lowell. But, despite personal wealth, the capital they controlled was inadequate to the task. More importantly, Lewiston was still at that time isolated geographically. The city was, in fact, a long way from seaboard shipping and the value of this region necessarily followed the expansion of railroads.

Locomotive traffic signaled the economic viability of Lewiston and it also launched out-of-state investment. Typically, the Lewiston Water Power Company (later reorganized as the Franklin Company) managed the most aggressive cotton factory development in Maine, building dams, canals, factories and machinery. Under the leadership of Boston capitalists, particularly Thomas Hill and Benjamin Bates, the systematic development of the city began at mid-century. In that time both Hill and Bates assumed the direct responsibility for several sites. The Bates Manufacturing Company began operation in 1852 and Lewiston Mills was added in 1853. In 1854 the Hill Manufacturing Company opened a new building while the Franklin Company developed its Lincoln Mill in the late 1850s as well. Androscoggin Mills was commenced in 1860 and by 1866 when the massive Continental Mills was completed the great period of construction was essentially over. Eight factories had been built with thirty-three water wheels turning a whopping 220,000 spindles. In the years after the Civil War the production of cotton goods reached thirty million yards per year with the employment of some 1,500 men and 3,500 women.

In terms of absolute production, Maine's experience was not extraordinary. Before 1810, Maine could point to only two cotton-spinning mills with a total of 780 spindles. Home weaving resulted in over 800,000 yards per year and over 1,000,000 yards of "mixed" fibers. Many yards were exported from Maine and, in relative terms, cotton manufacturing was emerging as one

This tintype of Delianne Aubin (left) and her friend, Madmoiselle Rousseau, was taken in Lewiston around 1880. *Courtesy Marietta Binette.*

This sterio card shows the spinning room of the Continental Mill, Lewiston, early in the twentieth century. *Courtesy Maine Historic Preservation Commission.*

of the most significant industries to the state's economy. On a per capita basis, even household production ranked well ahead of the national average in the early nineteenth century.

Wool

The manufacture of cotton cloth led directly to the creation of some of Maine's largest cities. But commonly, the manufacture of woolen goods did not. In both historical and technological terms the differences between the manufacture of cotton and wool were great. Machinery for carding and spinning was introduced for wool at about the same time it was developed for cotton. But cotton first came to New England in the form of machine-spun yarn while wool was raised on local farms and the processing of woolen cloth had been a domestic activity from the earliest years of settlement. The one, cotton manufacturing, drew the interest of speculators and investors. The other, woolen manufacturing, grew only slowly away from the home. Cotton factory cities were created in Maine but, with the exception of Sanford and Dexter, there were no woolen mill cities.

Carding machines and spinning jennies for woolen yarn were introduced in America in the 1780s at the same time as similar cotton manufacturing technology. The role of families in the use of wool tended to mitigate the impact of machinery. The initial use of this technology provided service to home spinners and weavers who often owned the sheep, wove the cloth and remained the true manufacturers of woolen cloth. The wool fibers were often a product of the home, and this manufacture was usually initiated by members of the family.

As a consequence of this orientation, Maine woolen manufacture did not lead to the birth of factory cities, the construction of boardinghouses or a vast immigration of labor until later in the nineteenth century, when the woolen factories of Dexter and Sanford were being created. While Maine's cotton factories were often large and few, woolen mills were smaller and more widely dispersed. Labor needs were the opposite of those found in cotton factory cities: in woolen factories some two-thirds of the employees were commonly men, as experience as well as physical strength was thought necessary to produce a variety of complex woolen products. Even as technology advanced from the jenny to the jack and power looms were introduced, the change in domestic manufacture was not dramatic. Aside from the work of carding mills, there was little evolution in woolen factories before the 1830s and up through the late 1820s there were only four woolen mills in the state, located in Gray, Lisbon, Gardiner and Bristol; all produced woven cloth.

The introduction of improved spinning machinery did little to make carding mills larger. If there was no extensive finishing at fulling mill establishments, the coarse woven material was flannel. When a design was woven into the face of the fabric, the result was cassimere, and when woolen yarns were used for filling and cotton was used for warps, the result was a combination of cotton and wool known as satinet. This latter was like the earlier linsey-woolsey and it was commonly woven on modified cotton looms produced from the late 1820s onward. This was the most common cloth woven in Maine woolen mills and in 1831 the state's four factories produced a total of 65,000 yards of satinet. This was a significant but not impressive quantity as compared to the continuing domestic manufacture. Early Maine mills were simply not part of a wider market economy and a number of combined cotton and woolen mills actually produced yarn for satinet. The decade of the 1830s, however, brought new technology and increasing specialization.

The Dexter Cassimere Factory was among the first in Maine to produce a product made entirely of wool. While only seven hundred yards of this was bound for Boston each month at first, this small start was an important turning point. The nearby Piscataquis Manufacturing Company in Dover-Foxcroft showed a market orientation by delivering its cassimeres overland to Bangor for ultimate sale in Philadelphia.

Throughout the nineteenth century, Maine was noted as being dotted with many small mill establishments and many of these, like Mayall's earlier factory in Gray, continued to provide services on toll for local residents. Many were very small. In 1831, for example, Dexter's cassimere factory employed only eight men and ten women and this was the largest woolen mill work force in the state! In Lisbon only twelve individuals ran the whole operation.

The number of woolen factories in Maine increased during the decade of the 1830s as new small mills were established in South Paris, Lewiston Falls, Sanford, South Berwick, Windham and Dixfield, as well as a number of other communities. Aside from custom work, most carding

The Libby mill in Lewiston was not the city's largest, but its prominent location near the Lewiston dam made it particularly attractive to photographers. *Courtesy Maine Historic Preservation Commission.*

was in support of cassimeres, satinet and broadcloth. Moses Greenleaf predicted in 1829 that "whenever Maine shall be distinguished in any considerable degree for its manufactures, that of wollen cloth, manufactured in the families of farmers, will form the principle article." In 1830 this had not happened in any measurable degree. But the manufacture of woolen goods was, indeed, destined to rank third in the value of the state's factory goods and, by mid-century, there were thirty-eight woolen factories in the state. Employment remained low, however: only 658 people, with the largest factory providing employment to less than 100.

Statistics lead to a premature conclusion that Maine's woolen industry was inconsequential. This is not true for a variety of reasons. Home manufacture was substantial and hard to measure, while factory-made goods accounted for some 3,700 yards per worker—a total of over 2 million yards per year. Perhaps most important, however, was the decentralized nature of the business

Erected in Lewiston, the Continental mill was relatively late but a handsome addition to the Lewiston skyline. *Courtesy Maine Historic Preservation Commission.*

and the relative small quantities of each "fancy" cloth marketed in the fashion industry. Rarely did woolen factories produce generic goods. It was more typical for them to produce small runs available to garment manufacturers as samples and made to order. Thus, the economic viability of very small mills was sustained as long as the quality and the designs suited the marketplace.

Certainly, the growth of Maine's woolen factories was slow. Of twenty-one woolen factories listed in 1860, fourteen would have been called very small: those in Limerick, York, Shapleigh, Alfred, Foxcroft, Lisbon, Windham, Mount Desert, Gardiner, Warren, Bethel, Hanover, Vassalboro and Hampden averaged only 14 workers. There were, of course, a few larger operations: a second mill in Vassalboro employed 180 while factories in Buxton, Bridgton, South Berwick and Lewiston retained as many as 80. For a period, the largest of the state's woolen factories was in Dexter where three related woolen mills were operated by R.W. Robinson. Here was, for the moment, the largest and most productive woolen factory complex in Maine with a staff of 200 men and women producing some 400,000 yards of cloth—more than all woolen factories in the state combined a decade earlier. There were located here some thirty carding machines, sixteen to twenty jacks and sixty power looms. Maine woolen factories were poised for a leap forward and, between 1860 and 1870, the number of factories rose from twenty-six to over one hundred, and employment increased to over 3,000.

Through the 1870s Maine's woolen manufactures grew apart increasingly from their carding mill forebears. After the Civil War, the national market economy provided the principal motive in establishing new factories. Around 1880, the Skowhegan Manufacturing Company, for example, wove between 800 and 1,000 yards per day using Crompton & Knowles looms and twenty weavers. Winter goods were made in the summer and summer goods in the winter. These were sent to Rufus S. Frost & Company of Boston, which then marketed the goods through their offices in Boston, New York and Chicago. By 1870 there were 107 woolen factories in Maine employing 1,582 men, 1,250 women and 210 girls. The largest concentration was in Dexter, where Amos Abbott & Company manufactured over 50,000 yards of cassimere and 17,000 yards of flannel.

The survival of small local factories ran counter to the prevailing spirit of the time, which often featured larger factories, bigger cities and cheaper labor. Centralization was manifested in the "buy-out" mentality, which fostered centralized control by assembling an archipelago of smaller factories into a larger enterprise. But it was discovered that a large factory size was no particular advantage. Thus, Maine factories became cogs in the larger wheel of the American Woolen Company.

Sanford

Most Maine woolen mills remained small, but there was one exception: the Goodall family enterprises in Sanford, started by Thomas Goodall in 1867. Here was the development of woolen and worsted mills that grew to be the largest such factories in Maine, attracting immigration that would rival many of the state's cotton factories.

Thomas Goodall had arrived in America for a short visit in 1846 from the center of British wool manufacture, Yorkshire. After nearly twenty years in both England and America, Goodall had purchased a mill site in Sanford and, typically, continued the manufacture of flannels at first and engaged in the production of relatively low cost goods.

It was estimated that between Springvale and Kennebunk, the Mousam River could produce nearly five thousand horsepower and most of this was wasted before the late 1860s. At Sanford alone, some ten thousand cubic feet of water per minute fell two hundred feet in a distance of several miles. This power was destined to transform the villages of Springvale and Sanford Corner. In 1869 it was noted that this waterpower "has been the making of Springvale, and is now building up Sanford Corner."

The success of Thomas Goodall and his sons resulted in the beginning from the production of specialty products commonly imported to America: fitted horse blankets and carriage robes. Both were finished consumer products sewn with finished borders, commonly printed with designs and often produced as plush fabrics. These were distinctions that defined the company's special place in the region's factory development.

Samuel Mayall built a multipurpose woolen factory in Gray in 1804, and the mill shown here was built by his daughter on the same property. He advertised a wide variety of services to households through the *Eastern Argus* in Portland. Primarily these services were offered on a bespoke basis. *Courtesy Maine Historic Preservation Commission.*

Since many products were marketed as finished pieces, sewing machines were needed to create sizes and form borders. Although woolens were not usually printed, block printing was needed to create a variety of horse blankets and carriage robes, sometimes using cotton warps as well as woolen filling. Looms were employed, which used fabric stretching and knives to cut raised pile into plush fabric in a way that was similar to the making of corduroy or Bristol-type carpets. The company used double-shuttle plush looms to produce this specialty fabric and mohair car and furniture plushes were made starting in 1881.

Of particular importance was the production of worsted, beginning in 1889, and. much later, the manufacture of Palm Beach suiting. Both the production of woolens and worsted cloth began with the raw material of wool. For this reason, both of these products have been frequently lumped together. In reality, though, each has experienced a different history and different technology. In short, it can be said that woolens begin with the teasing of curly, barbed fibers of wool so as to produce a uniform, carded rolag suitable for spinning. In the case of worsted, however, wool fibers are separated by length and combed to make possible lighter uniform yarn similar to that used in the cotton industry and suitable for the manufacture of men's suits and uniforms in addition to finer women's clothing. The Goodall Worsted Mill allowed for the production of both fine upholstery and suiting using worsted spun in Sanford and nearby Limerick.

This small woolen mill built in Gray in 1834 produced a variety of fancy cloths by power loom. *Courtesy Maine Historic Preservation Commission.*

Shoemaking

In larger American cities, the factory-like organization of shoemaking began as early as the mid-eighteenth century. In Maine, though, this manufacture continued to result from the work of itinerant craftsmen or local cobblers who sometimes used leather provided by the clients. Traditional pattern continued into the nineteenth century and local cobblers' shops were common until modern times. In an age of ever-increasing specialization, the custom manufacture of shoes declined in the face of regional and even national trade. Increasingly the shoe emporium was the local general store. As late as 1852 the store of Owen Doyle inn Benedicta, for instance, included some 260 pairs of boots and shoes. In Gardiner, the downtown storefront of Michael Burke offered 850 pairs.

The evolution from workshop to factory production methods was accelerating by 1870. As early as 1850 the state had counted some 3,500 shoemakers, but this count illustrated no particular chronology of evolution and the total included workbench trades carried on in homes, small workshops and occasionally small factories employing over 100 people. True factories emerged from this mixture and the factory development of shoemaking in the city of Auburn was especially notable. During the second half of the nineteenth century, Auburn became the state's shoemaking center and one of the most important shoemaking centers in New England. After the Civil War the annual production of shoes

Workers posed in front of the woolen mill built by local farmers in East Wilton around 1842. *Courtesy Maine Historic Preservation Commission.*

occupied 3,000 to 4,000 employees in twenty-one different factories producing a total of nearly four million pairs of shoes.

The job of the eighteenth-century cordwainer was largely redefined in factory terms during the last third of the nineteenth century as a result of mechanization, division of labor and product specialization. The manufacture of women's shoes was centralized in eastern Massachusetts, particularly Lynn, and this manufacture was so well centralized that few Maine companies sought to compete for this market. Still, in 1830, Balch & Furbish of Portland produced "Ladies figured silk Boots" and "Ladies figured silver bronze slippers" as well as men's shoes. At mid-century, J.J. Davis and Son manufactured "Ladies Gaiters and Fancy Shoes," and the nearby firm of Jonas Tibbetts of Saco employed twenty men and twenty women to produce more than twenty-one thousand pairs of ladies' shoes each year. More typical of many Maine shoemakers, though, were such items as boots and thick boots, as well as brogans (ankle-high cowhide shoes with thick soles). Machinery substituted for craft skills and fostered the development of these industries in regions of lower-cost labor. Stitching machines, especially the McKay Stitcher, could be used to attach the uppers to soles. A large and uncomfortable welt was left interior to the shoe, however, and most of these first mass-produced men's shoes were sold to the South as low-cost goods.

The process of shoemaking was quite similar everywhere. When in homes, shops or manufactories, the first step was the tanning of leather, though this was most frequently a

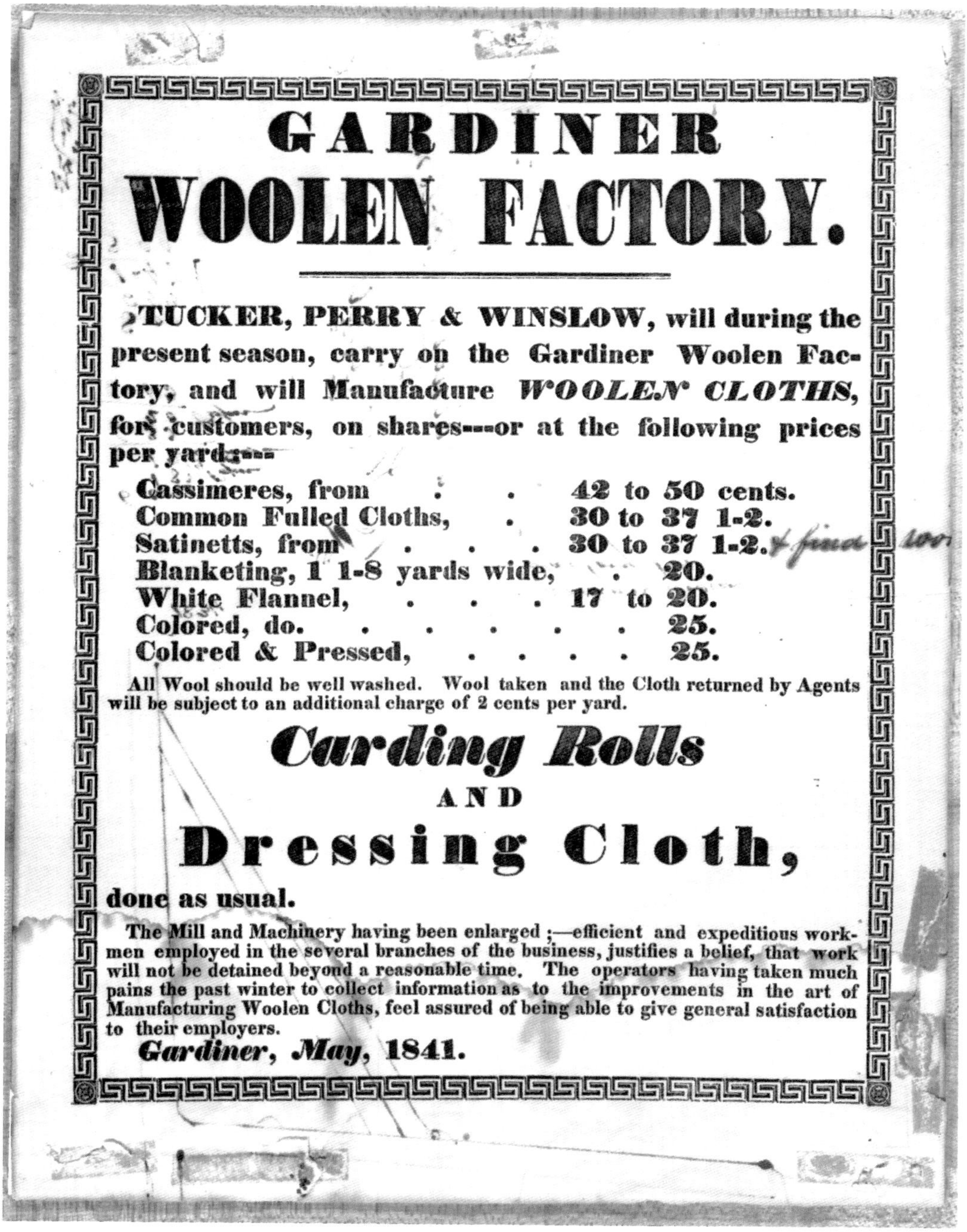

GARDINER WOOLEN FACTORY.

TUCKER, PERRY & WINSLOW, will during the present season, carry on the Gardiner Woolen Factory, and will Manufacture *WOOLEN CLOTHS*, for customers, on shares---or at the following prices per yard:---

Cassimeres, from . .	42 to 50 cents.
Common Fulled Cloths, .	30 to 37 1-2.
Satinetts, from	30 to 37 1-2.
Blanketing, 1 1-8 yards wide, .	20.
White Flannel, ...	17 to 20.
Colored, do.	25.
Colored & Pressed, ...	25.

All Wool should be well washed. Wool taken and the Cloth returned by Agents will be subject to an additional charge of 2 cents per yard.

Carding Rolls

AND

Dressing Cloth,

done as usual.

The Mill and Machinery having been enlarged ;—efficient and expeditious workmen employed in the several branches of the business, justifies a belief, that work will not be detained beyond a reasonable time. The operators having taken much pains the past winter to collect information as to the improvements in the art of Manufacturing Woolen Cloths, feel assured of being able to give general satisfaction to their employers.

Gardiner, May, 1841.

As with many Maine woolen mills in the early 1840s, the Gardiner Woolen Factory produced a variety of cloths using wool often provided locally. Satinet, using both cotton and wool, was the most common product of these mills. *Courtesy Maine State Museum.*

A sample of woolen yarn can still be seen on this jack in use within a small Maine woolen factory around 1910. *Courtesy Maine State Museum.*

separate enterprise. Cutting and sizing were the first workshop steps, followed by sewing the uppers and tacking them temporarily to lasts (wooden feet).

The development of shoemaking factories was nevertheless unique in one characteristic—as mechanization progressed, patents and machinery production were centralized in the manufacture and distribution network of the United Shoe Machinery Corporation of Beverly, Massachusetts. United controlled the market for machinery and they did not sell it—they leased it. Thus, investors in shoe factories were required to invest less capital and, accordingly, Maine was soon filled with many shoe shops using leased machinery and buildings with reduced tax burdens to create local jobs.

Engineering

Among the prevailing characteristics of most nineteenth-century manufactures was the inexorable replacement of craft processes by machinery of all sorts. The century was indeed dominated by the spread of machines and engines. In turn, both resulted from and promoted the "factory system" of manufactures. Steam power, perfected in the eighteenth century, became a force in the nineteenth. This fostered industrial expansion in remote regions, revolutionized transportation on both land and sea and forever altered the demographic distribution of industrial development of America. In Maine, as elsewhere, the early decades of the nineteenth century ushered in an era

A bank note produced by the Sanford Bank in the 1860s features a stock engraving of power loom weaving. *Courtesy Maine State Museum*.

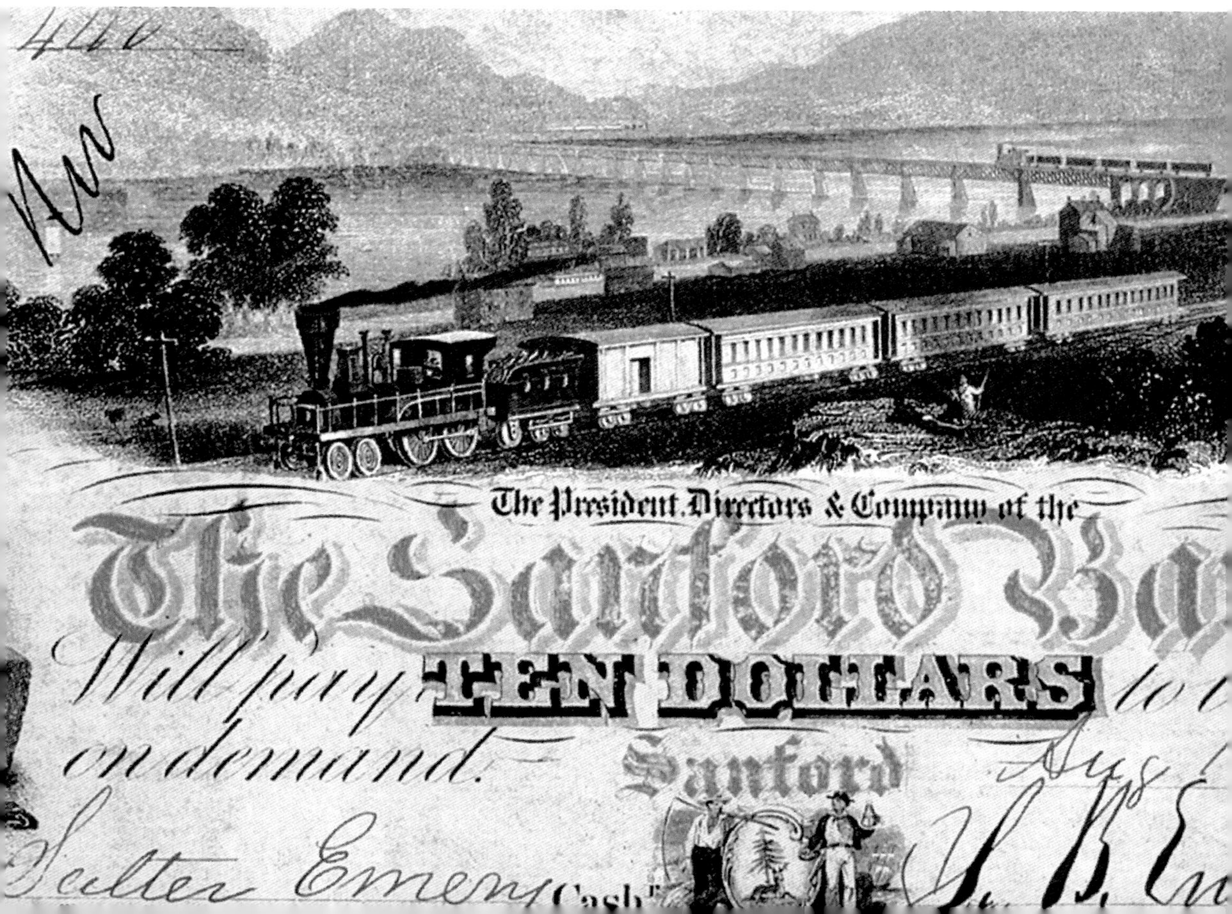

This Sanford bank note includes the signature of a prominent local merchant and banker, S.B. Emery. *Courtesy Maine State Museum*.

This view of the Cascade woolen mill probably dates from the 1880s. Reflected here, however, was a style of mill architecture more characteristic of the 1840s. *Courtesy Maine Historic Preservation Commission.*

Weaver Ken Daggett poses here between two Crompton & Knowles looms at the woolen mill in North Monmouth in 1918. *Courtesy Maine State Museum.*

of even larger foundry and machine shop industries, starting to produce both cotton and woolen machinery, boilers and steam engines to drive both factories and steamboats, machine tools designed to make other tools and machines and locomotive engines to serve the nation's rising rail system. These nineteenth-century developments brought a new era of engineering and machine building, which expanded and changed the roles of both machine shops and foundries.

Foundries seldom produced consumer products and even tinsmiths were generally the true manufactures of stoves using cast stove plates. When foundries were combined with machine shops, however, the stage was set for the manufacture of consumer goods and this was certainly the case with clocks, firearms, sewing and knitting machines. The largest foundry/machine shops manufactures, however, served the needs of one group of manufacturers, or even just one manufacturer. Again, these industries were removed from individuals and so they were commonly subordinate and subsidiary to cotton and woolen manufacturers, railroads and factories of all sorts.

Analysis of Maine's nineteenth-century industrial manufactures can easily be hopelessly complicated. Viewed from a technical perspective, several characteristics can be isolated: the shift from wood to iron and steel; the development of precision machine tools; and the emergence of machine and engine building as distinct branches of manufacturing.

When most machines were made primarily of wood, the role of the foundry was secondary, and millwrights showed a prejudice toward woodworking. The construction of light castings advanced the use of iron and steel components as fasteners and these, as well as machinery parts, required greater care than the casting of pots and stove parts. Production of cast water turbines in Garland, Skowhegan, Gardiner and Bangor fostered this evolution.

A second characteristic of nineteenth-century machine building was the development of precision tools. Up to the end of the eighteenth century, the products of most nineteenth-century manufactures were still being custom made. While the concept of mass-produced parts had been established long before in the manufacture of Windsor chairs, these manufactures did not require any great precision. Clocks, firearms and spinning machines, by contrast, demanded identical, and therefore interchangeable, parts. This was not a task for the woodworkers or for the blacksmith either. Rather this new trade of machinist was born—individuals who, through the cutting, turning and grinding of cold metal, could dispense with custom crafts in favor of mass-produced uniformity. Pioneers in the development of precision machine tools included textile machinery builders.

Tool making was historically ancillary to the work of craftsmen who, as a matter of need, often made their own tools. A third characteristic of nineteenth-century machine building was the emergence of machinery as a separate branch of manufacturing. This development was retarded by the tradition of mill construction in which machines could not, in fact, be separated from mill structures themselves. The mechanisms of sawmills and the grindstones of gristmills were often connected directly to water wheels and were not portable—they were the work of millwrights, not machine shops.

Textile factories introduced factory-type organization to the region, and at the same time they also introduced the manufacture of machinery and machine technology. Machine building had not yet emerged as a distinct factory task and, consequently, the job of building the machinery and selling it to others was a secondary consideration. The machine shops

This postcard view of a shoe shop in Belfast dates from around 1900. *Courtesy Maine Historic Preservation Commission.*

assembled by textile mills had access to the best machinery and the most experienced labor pool of machinists. It is no surprise, then, that textile machine shops led the nation in engineering. When, for example, the design of the USS *Monitor* required the machining of the track to rotate the turret, this critical job was handled where it could be done best: the Lowell machine shop.

Machinery

Engineering and machine-building industries emerged as major employers and an important part of the state's economy. But the pattern of these products remained task-oriented rather than market-driven. The Saco machine shop at first worked for the parent company on a jobbing basis.

The first factories of great size were those for spinning and weaving cotton and woolen cloth and these enterprises commonly began with the construction of the machinery they would need. Machine shops were commonly started as soon as the first floor of the mill was ready. The example provided by Saco is compelling. Here the waterpower on Cutts Island was transformed from the manufacture of boards and nails beginning in 1826. The Saco Manufacturing Company was formed to make cotton cloth but this was not possible until machines could be built. Some 735 machines were needed, including 224 spinning machines, 300 looms and 120 carding machines. This was a tall order, but the Saco shops managed to complete over 400 of these machines before the large wooden factory itself was burned.

The work of the Saco machine shops was necessarily tied directly to the needs of the parent company. Operation followed a traditional pattern of master and apprentice, which was unlike

Through most periods up to the mid-twentieth century, the manufacture of rope and twine demanded ropewalks, long buildings like the example shown here in Castine, circa 1900. *Courtesy Maine Historic Preservation Commission.*

the pattern found later in the cotton factory itself. Considerable craft skill was still required to finish parts, and some forty-four bench vices were listed in the inventory of 1829. Machine building included an employer/employee relationship with contracts awarded to the principals for the manufacture of machines or machine parts, not employees.

Between 1826 and 1850 the work of Saco machine shops was exclusively to fill orders from Saco and Biddeford companies. After mid-century, though, there was an effort to specialize in the production of only certain cotton machines. After 1856 the Saco Water Power Company also offered mill gearing, some machine tools and Dowd, Bradbury and Spiral-Vent water wheels, as well as sawmill work. In the 1860s the Saco shops came to specialize in the production of speeders and fly frames used in the production of cotton roving, though as late as 1886 they produced a full line of machinery for the Edwards factory in Augusta.

Employment in the Saco shops varied wildly depending on contracts and on the overall economy. While in 1832 there had been around one hundred machinists at work, the number of machinists and helpers occasionally reached six hundred in the 1840s and 1850s. Here was a source of needed machinery and also a training ground for a whole generation of machinists.

Not surprisingly, a number of foundry/machine shops turned to the needs of the state's lumbering enterprises, producing sawmills, planers and other tools. The T.H. Ricker company of Harrison was the largest; mill gearing and steam engines were made by the Penny company of Mechanic Falls; Lombard Log Haulers were made by the A.O. Lombard Company of Waterville. In Lewiston, the Lewiston Machine Company employed some 180 people to produce machinery of all sorts. Meanwhile, the firm of Fay and Scott of Dexter became the state's premier manufacturer of machine tools with a large offering of lathes and planers, while shops along the coast turned out windlasses, pumps and other marine commodities used in shipbuilding.

This is an exterior view of shoe workers outside the M.M. Richard clothing factory in Gardiner, circa 1910. *Courtesy Maine Historic Preservation Commission.*

A workman rests at the Lombard factory in Waterville, around 1900. *Courtesy Maine State Museum.*

Throughout the century, however, machine builders and engineering companies remained ancillary to other interests, be they in the textile factories of Saco, Biddeford, Lewiston, Dexter and Sanford, or in the engines needed to run sawmills or inland factories. Maine remained dominated by the needs of the forest industries. Not until the end of the nineteenth century did the advance of iron foundries and steam engines lead to the Bath Ironworks. The triumph of iron and steam over wind and wood had taken a full century.

Engines

The task orientation of engine-building was like the work of shipyards, and this special order characteristic restrained these industries from full development of a factory system. These enterprises tended to remain organized as shops even while their large capitalization and division of labor suggested them as factories. Locomotive engines always tended to be like ships. They were given names, and lithographs were produced to celebrate them. Each was a grand engineering achievement.

While machine-building in Saco was promoted by the cotton industry, the state's largest nineteenth-century foundry/machine shop, the Portland Company, arose from the desire to build a railroad from Portland to Montreal (the Atlantic and St. Lawrence Railroad) and, consequently, the need for locomotive engines.

Waterpower and sailing ships were easily identified, so steam power was slow to take root in Maine. Yet between 1846 and 1892, the Portland Company was destined to produce some nine hundred locomotive engines, as well as other engines, fire pumpers and machinery. The Portland Company was organized in 1846 expressly to build locomotives and cars for the railroad from Portland to Montreal. These engines were built on a wider track than normal to prevent the connection of other lines shortcutting the route to Boston. This custom construction proved shortsighted, and the strategy did not make the line successful. As a byproduct, though, the Portland Company grew to become the state's most substantial engineering and machine-building establishment. During just the first three years of operation, the company produced ten locomotive engines, nine passenger cars, thirty platform cars and forty freight or boxcars. Like the Saco shops, there was, at first, only one customer for all this material, but activities here took on a life of their own. The Atlantic and St. Lawrence line was a failure, but the Portland Company was a notable success.

By mid-century the Portland Company was producing some five locomotives each year and 150 machinists and laborers were employed out of a total pool of machinists in Maine of only 500. Here the work of the foundry and machine shops were combined in a way unprecedented in Maine: the products transcended local needs and included marine engine boilers, snow plows, fire engines and, ultimately in the twentieth century, the manufacture of automobiles.

Following the Civil War, during which the Portland Company produced ordnance of various types, the company easily became the state's largest foundry/machine shop. In 1870 it boasted the iron foundry employing thirty-five men and a brass foundry and "machine shop boiler shop and car factory" employing nearly two hundred men. During these years, this company produced thirteen locomotives, fifty-one cars, other engines and tools and one fire engine (Cumberland Steamer No. 2) for Portland. Inside were located no less than fifty-five lathes, twenty-nine planers, sixteen drills and over fifty other major machine tools.

The Portland Company was unrivaled in nineteenth-century Maine, but it was certainly not alone. In 1850 the small Portland company of Winslow & Crockett produced four steam engines and five boilers with a staff of nine. Nearby, Daniel Freeman produced two steam engines and boilers and, in Bangor, Hinckley & Egery made water wheels, mill gearing, steam engines and boilers. By 1870 some seventy men were employed here to produce both mill machinery and steam engines.

Social Changes

Factory manufactures, particularly textiles, brought great economic and social changes to Maine. The economic changes were noticeable right away—mechanization helped move manufacture from the home, worked to regiment and centralize manufacturing and created towns and cities in the place of pastures and farmhouses. Fundamental social changes inexorably followed.

At first textile factories copied the social order they saw working in Rhode Island and Massachusetts. In Rhode Island cotton-spinning mills had first used children for the majority of work—the sons and daughters of neighbors and investors. But Maine factories most often followed the model established in Waltham and, later, Lowell. Here the bulk of the factory's work was managed by young women living in company boardinghouses and drawn from the pool of Yankee farmers' daughters in New England. Spirited and independent, these girls were anxious for the chance to be on their own and make their own money. But they were not yet independent enough to argue out loud. Accordingly, the labor of Maine's first cotton factory cities were as they imagined Lowell to be.

While work in the home declined, the role of women in the cotton industry was of critical importance and women dominated this labor pool. In the 1830s, the factory in Cumberland employed 15 men and 57 women, the Gardiner factory had 15 men and 72 women and the Winthrop factory offered employment to only 12 men and 83 women. But the era of these small cotton mills, and others in North Berwick, Oxford and Augusta, was nearly over, replaced between 1830 and 1840 by Saco and Biddeford. Following the lead of Lowell, the workforce assembled on Factory Island was decidedly female. And most girls and women lived at home, in local families, or more generally, in housing units adjacent to the factories. The units, described variously as tenements, boardinghouses or blocks, were usually supervised by other mature women, housemothers protecting the daughters of farm families from the evils of urban life. In 1840, for example, there were 26 housemothers managing a flock of some 222 women between the ages of fifteen and twenty and 264 between twenty and thirty. Mrs. Day, for instance, operated a boarding house of 56 young women. Commonly, there was one matron for every 12 to 25 women. With the start of Laconia and Pepperell factories in Biddeford, the same system was spread there. At first these were Yankee girls, transient workers who fully expected to return to their homes. Many sought a taste of the world, not a life to live. Beginning in the 1850s this workforce became somewhat more difficult to control (they had often been driven from their hometowns by poverty and a lack of opportunities) and they began to be replaced by Irish immigrants happy to find cash wages.

Investors in Lewiston followed the same labor pattern previously used in the Saco/Biddeford area, even though this was already changing. Boardinghouses arose along with the factories themselves, and the city soon became a town of many boardinghouses. In 1870 there were some

PASSENGER
WITH
IMPROVED VARIAB
Diam. of Cylinder
Length of Stroke
Diameter of Driving
Total weight of Engine
PORTLAND COS L
James C. Churchill Agent
PORTLA

The *Forest State*, a passenger engine, was just one of some nine hundred locomotives built by the Portland Company from the mid-1840s through the mid-1890s. *Courtesy Maine State Museum.*

full-time boardinghouse keepers like Thomas Day who, with his wife, managed accommodations for 1 man and 37 women. The Bates Mill houses were like those of Sarah Fogg, which included 2 men and 19 women, or those of Emeline Hussey and Esther Fitzgerald, whose houses were slightly larger. A campus of sorts for young women arose near the Bates factory. The largest boardinghouse was run by Josiah Flint, whose tenants included three complete families, 115 women and 44 men. It is important to note that until the 1870s most boarders were of Yankee extraction, following the Lowell pattern, and most were Maine natives. But in the early part of the century, this was already beginning to change in a dramatic way.

Until textile manufacture blossomed in New England, the most common job for young women was work as domestics in the larger households of the more wealthy. This placement was perhaps popular among the heads of poorer families, but it appears to have been unpopular among the women themselves. When given the choice of minding spinning machines instead of children and household duties, the exodus to factory life, and cash wages, began. But the attraction of factory cities soon began to fade as expansion, competition and depressed wages undermined the once higher quality of life in these cities. Beginning as early as the 1840s, Yankee women went home and stayed there, their places now increasingly taken by poorer immigrants.

The Irish were the first to arrive in the late 1840s and they soon took the vacated jobs of domestic servants and soon textile factory operatives. There was resistance to this, but in Maine the impact of this immigration was limited. Displacement came later in the 1860s, and particularly the 1870s and 1880s, with the arrival of French Canadians from Quebec. From that point, the development of textile factory cities and French Canadian immigration were tied together inexorably in New England.

As factory jobs became less desirable, and as the agricultural base of Quebec disintegrated, French families from eastern Canada were sought to provide new sources of cheap labor for New England factories. As a direct consequence, French became the dominant subculture of Maine as French immigrants permanently changed the demography of Saco/Biddeford, Lewiston, Sanford, Augusta, Waterville and, in fact, any other community engaged in the manufacture of textiles. Descendants of these workers still form the populations of many of these communities today.

Emmigrants from Quebec were unique in American history: this was a population that came to New England cities by train. Home was nearby, just a one-way ticket back to the farm. Accordingly, many did not see relocation to New England as permanent and there was little incentive to assimilate within the Yankee society. The move was often one of colonization, not immigration. As whole towns emptied out, these French Canadians brought with them their language and their religion and they were not prepared to surrender either. Segregating themselves, the "Little Canadas" of New England were lumps in the cream of wheat.

When factory labor was dominated by young women, the flexibility of this employment brought some independence. But the arrival of French from Quebec meant the wholesale relocation of communities. Entire families now worked in the factories and everyone required employment to make ends meet. This was a dangerous economic necessity and dependence on the mill grew here. New arrivals effectively ended dormitory life and the boardinghouse system. It suggested the beginnings, instead, of the tenement house. This was a return to the mill village of Rhode Island of fifty years before, but this time with the bonds of a common cultural language and religion. Here was a return to child labor, with its consequences excused by the cultural divide. The workforce, the demography and the economy of Maine were changing, and not particularly for the better. In places like Biddeford and Lewiston it was quickly seen that only the children of immigrant families were employable in the factories. Commonly, the heads of the households became laborers while the wife, with small children still in her care, was at home, and numerous children from ages eight to eighteen actually secured wages at the factory. The family of Joseph Gilbert was typical. Joseph, thirty-seven, was a laborer. His wife tended four small children under the age of seven. An older disabled child remained at home but Mary, sixteen, Joseph, fourteen, Eliza, twelve, and Anna, eight, were employed in the mills. A decade later the pattern was still the same: Felix Bisson, fifty-one, was a laborer and his wife was engaged in housekeeping. Two daughters, eighteen and twenty-five, and three sons, ten, twelve and fourteen, worked in the mills.

The textile factories emerging along Maine rivers revolutionized the ethnic composition and economy of the state, a fact that has held true to the present day.

Knox

This circa 1910 photograph shows a motorcar recently assembled in Portland and found in the files of the Portland Company. *Courtesy Maine State Museum.*

Notes on Sources

This book was originally intended as a companion to the major Maine State Museum exhibit of the same name. The first draft was researched at the same time the exhibit was planned, from 1979 to 1985. During this time I worked on the installation daily and worked on this book at home during the evenings. I seldom was away from the museum during these years and, indeed, there was no need for me to be away, since virtually all of the essential material was available in Augusta, and most in the Cultural Building where the exhibit was mounted.

From the Maine State Museum I secured much information using its resource center. This provided me with a great deal of important data and gave me access to papers from the first two years of the *Scientific American.* These resources were used for both the exhibit and this book.

Two other major Maine organizations were also located near the museum. At the Maine State Library I found local histories of most Maine cities and towns. These were variable in quality, but I looked at them all and used many. The library also provided microfilm copies of many nineteenth-century Maine newspapers, including advertisements posted by the proprietors of early state businesses.

Also located near the museum entrance was the Maine State Archives. The archives provided microfilm of census records, including the all-important 1850 Industrial Census. Also at the archives were a copy of the diary of carpenter Moses Davis as well as town records from Gorham detailing the history of that town producing carpets.

Little was needed outside the Cultural Building, but of particular help was the important collection of the Maine Historic Preservation Commission. In the building of the exhibit, and the publication of this book, this photo collection has proven to be an irreplaceable resource.

Most writing for this was done at home in Hallowell and on frequent field trips that took me to the sites of Maine enterprises, including Samuel Mayall's mill in Gray.

Index

Selected List of Maine Museums

The state of Maine offers more than two hundred historical societies and museums of all kinds and sizes, and many include local history libraries where readers can study local manufacturers and industries. A selected few of these are listed below. Readers may also wish to visit local public libraries for more information on the history of manufacturing in the Pine Tree State.

Acadian Village
RR 1
VanBuren, Maine 04785

Bangor Museum and Center for History
159 Union Street
Bangor, Maine 04401

Brick Store Museum
117 Maine Street
Kennebunk, Maine 04043

Camden-Rockport Historical Society
Conway Road
Camden, Maine 04843

Franco-American Heritage Center
46 Cedar Street
Lewiston, Maine 04241

Lumberman's Museum, Inc.
61 Shin Pond Road
Patten, Maine 04765

Maine Historical Society
489 Congress Street
Portland, Maine 04101

Maine Maritime Museum
243 Washington Street
Bath, Maine 04530

Maine State Museum
State House Complex
Augusta, Maine 04333

Old Fort Western
16 Cony Street
Augusta, Maine 04330

Old York Historical Society
207 York Street
York, Maine 03909

Pejebscot Historical Society
159 Park Row
Brunswick, Maine 04011

Penobscot Marine Museum
5 Church Street
Searsport, Maine 04974

Portland Museum of Art
7 Congress Square
Portland, Maine 04101

Robert Abbe Museum
Sier de Monts Spring
Acadia National Park
Bar Harbor, Maine 04609

Relevant Museums Outside of Maine

Mystic Seaport
75 Greenmanville Avenue
Mystic, Connecticut 06355

Old Sturbridge Village
1 Old Sturbridge Village Road
Sturbridge, Massachusetts 01566

Strawbery Banke Museum
Marcy Street
Portsmouth, New Hampshire 03801